Intelligent and Connected Vehicle Security

RIVER PUBLISHERS SERIES IN SECURITY AND DIGITAL FORENSICS

Indexing: All books published in this series are submitted to the Web of Science Book Citation Index (BkCI), to SCOPUS, to CrossRef and to Google Scholar for evaluation and indexing.

The "River Publishers Series in Security and Digital Forensics" is a series of comprehensive academic and professional books which focus on the theory and applications of Cyber Security, including Data Security, Mobile and Network Security, Cryptography and Digital Forensics. Topics in Prevention and Threat Management are also included in the scope of the book series, as are general business Standards in this domain.

Books published in the series include research monographs, edited volumes, handbooks and textbooks. The books provide professionals, researchers, educators, and advanced students in the field with an invaluable insight into the latest research and developments.

Topics covered in the series include, but are by no means restricted to the following:

- Cyber Security
- Digital Forensics
- Cryptography
- Blockchain
- IoT Security
- Network Security
- Mobile Security
- Data and App Security
- Threat Management
- Standardization
- Privacy
- Software Security
- Hardware Security

For a list of other books in this series, visit www.riverpublishers.com

Intelligent and Connected Vehicle Security

Jiajia Liu
Northwestern Polytechnical University, China
Abderrahim Benslimane
Avignon University, France

Published 2021 by River Publishers
River Publishers
Alsbjergvej 10, 9260 Gistrup, Denmark
www.riverpublishers.com

Distributed exclusively by Routledge
4 Park Square, Milton Park, Abingdon, Oxon OX14 4RN
605 Third Avenue, New York, NY 10017, USA

Intelligent and Connected Vehicle Security / by Jiajia Liu, Abderrahim Benslimane.

ISBN 978-87-7022-367-6 (print)

Contents

Preface xi

List of Figures xiii

List of Tables xix

List of Abbreviations xxi

1 Vehicle Bus Security **1**

1.1 Vehicle Bus 1

1.1.1 Overview of Vehicle Bus 1

1.1.2 Categories of Vehicle Bus 2

1.1.2.1 LIN Bus 3

1.1.2.2 CAN Bus 5

1.1.2.3 FlexRay Bus 8

1.1.2.4 MOST Bus 9

1.1.2.5 Automotive Ethernet Bus 11

1.1.3 Threat Analysis of Vehicle Bus 12

1.2 CAN Bus Vulnerability and Analysis 13

1.2.1 CAN Bus Architecture 13

1.2.2 On-Board CAN Bus Access 17

1.2.3 Reverse CAN Bus Communication 19

1.2.4 Analysis of CAN Message Data 21

1.2.5 Fuzz Testing of CAN Bus 22

1.3 Analysis of OBD-II Interface Attack Technology 24

1.3.1 Overview of OBD-II 24

1.3.2 Attack Technique Analysis of OBD Box 26

1.3.2.1 Attack Surface Analysis 27

1.3.2.2 Attack Process Analysis 29

1.3.2.3 Invade Car CAN Bus 35

1.4 Attack Experiment Against Vehicle Bus 35

1.4.1 Control of the Display of Vehicle Dashboard 36
1.4.2 Vehicle Status Tampering via Body Control Module 39
1.4.2.1 Control the Door Lock Switch 40
1.4.2.2 Car Lights and Wiper Attacks 41
1.4.3 CAN Bus Overload via Flood Attack 42
1.5 Experiment of Driving Behaviour Analysis 43
1.5.1 A Novel Scheme About Driving Behaviour Extraction 43
1.5.1.1 Extraction and Processing of Driving Data 44
1.5.1.2 Grade Regulation and Label Design 45
1.5.1.3 Establishment of MTL Network 46
1.5.2 Experimental and Numerical Results 47
1.5.2.1 Volunteer Recruitment and Experiment Route 47
1.5.2.2 Illegal Driver Detection Results 47
1.5.2.3 Legal Driver Identification Results 49
1.5.2.4 Driving Behaviour Evaluation Results . . 49

2 Intra-Vehicle Communication Security 53
2.1 Basic Introduction to Intra-Vehicle Communication 53
2.1.1 Overview of Intra-Vehicle Communication 53
2.1.2 Module Threat Analysis 54
2.1.2.1 Threat Analysis of Keyless Entry 54
2.1.2.2 Threat Analysis of TPMS 54
2.1.2.3 Threat Analysis of Wi-Fi 55
2.1.2.4 Threat Analysis of Bluetooth 55
2.1.2.5 Threat Analysis of FM 57
2.1.2.6 Threat Analysis of GPS 57
2.2 RKE/PKE Security 58
2.2.1 Overview of RKE 58
2.2.2 Overview of PKE 60
2.2.3 Attack Technique Analysis of RKE/PKE 62
2.2.3.1 Security Technology of RKE/PKE System 62
2.2.3.2 Summary of Common Attacks of RKE/PKE System 64
2.3 TPMS Security 66
2.3.1 Overview of TPMS 66
2.3.2 Attack Technique Analysis of TPMS 70
2.4 On-Board Wi-Fi Security 72

2.4.1 Overview of On-Board Wi-Fi 72
2.4.2 Attack Technique Analysis of On-Board Wi-Fi . . . 74
2.4.2.1 WEP Protocol 74
2.4.2.2 WPA/WPA2 Protocol 76
2.4.2.3 WPA3 Protocol 81
2.5 On-Board Bluetooth Security 82
2.5.1 Overview of On-Board Bluetooth 82
2.5.2 Attack Technique Analysis of On-Board Bluetooth . 84
2.5.2.1 Status of Bluetooth Security Mechanism . 86
2.6 FM Security 91
2.6.1 Overview of Radio 91
2.6.2 Attack Technique Analysis of FM 93
2.6.2.1 Introduction of FM Radio 94
2.6.2.2 Attack Process of FM Radio 94
2.7 GPS Security 96
2.7.1 Overview of GPS 96
2.7.2 Attack Technique Analysis of GPS 99
2.8 Experiments 101
2.8.1 Attack Experiment Against RKE 101
2.8.1.1 Experimental Environment 101
2.8.1.2 Experimental Method 102
2.8.2 Attack Experiment Against TSP 103
2.8.2.1 TSP Attack Process 104
2.8.2.2 Experimental Results of TSP Attack . . . 105
2.8.3 Attack Experiment Against TPMS 106
2.8.3.1 Structure of TPMS 107
2.8.3.2 Reverse Engineering TPMS Communication Protocols 108
2.8.4 Attack Experiment Against Vehicle Bluetooth via BlueBorne Vulnerability 109
2.8.4.1 What is BlueBorne? 110
2.8.4.2 Process of BlueBorne 111
2.8.5 Attack Experiment Against Road Navigation System 111
2.8.5.1 Goals of Road Navigation System 112
2.8.5.2 Detailed Attacking Process 113
2.8.5.3 Experiments and Results 114

3 Unmanned Driving Security and Navigation Deception 117
3.1 Basic Introduction to Unmanned Driving 117

3.1.1 What is an Unmanned Vehicle? 117
3.1.2 Core Functional Modules of Unmanned Driving . . 118
3.1.2.1 Perception 119
3.1.2.2 Planning 121
3.1.2.3 Control 123
3.2 Ultrasonic Radar Security 126
3.2.1 Overview of Ultrasonic Radar 126
3.2.2 Basic Principle of Ultrasonic Radar 130
3.2.3 Attack Technique Analysis of Ultrasonic Radar . . . 131
3.2.3.1 Jamming Attack 132
3.2.3.2 Spoofing Attack 133
3.3 Millimeter Wave Radar Security 133
3.3.1 Overview of Millimeter Wave Radar 133
3.3.2 Basic Principle of Millimeter Wave Radar 135
3.3.2.1 Ranging Principle 136
3.3.2.2 Velocity Measuring Principle 137
3.3.2.3 Target Recognition 138
3.3.3 Attack Technique Analysis of Millimeter Wave Radar 138
3.3.3.1 Jamming Attack 139
3.3.3.2 Spoofing Attack 139
3.3.3.3 Relay Attack 140
3.4 Hi-Definition Camera Security 140
3.4.1 Overview of Hi-Definition Camera 140
3.4.2 Basic Principle of Hi-Definition Camera 143
3.4.3 Attack Technique Analysis of Hi-Definition Camera 144
3.4.3.1 Blinding Attack 145
3.5 LiDAR Security 145
3.5.1 Overview of LiDAR 145
3.5.2 Basic Principle of LiDAR 147
3.5.3 Attack Technique Analysis of LiDAR 151
3.5.3.1 Relay Attack 151
3.5.3.2 Spoofing Attack 152
3.6 Experiments 152
3.6.1 Attack Experiment Against Ultrasonic Radar 152
3.6.1.1 Jamming Attack 153
3.6.1.2 Spoofing Attack 154
3.6.2 Attack Experiment Against Millimeter Wave Radar 155

3.6.2.1 Jamming Attack 155
3.6.3 Attack Experiment Against Hi-Definition Camera . . 157
3.6.3.1 Blinding Attack 157
3.6.4 Attack Experiment Against LiDAR 160
3.6.4.1 Relay Attack 160
3.6.4.2 Spoofing Attack 162

Bibliography **167**

Index **169**

About the Authors **171**

Preface

Intelligent and Connected Vehicles (ICVs) are becoming the mainstream of worldwide future automotive industry. A lot of advanced technologies, like artificial intelligence, big data, millimeter wave radar, LiDAR and high-definition camera based real-time environmental perception, etc., are increasingly being applied towards the ICVs, making them more intelligent and connected with things surrounding the vehicles. On the other hand, the mobile communication and networking techniques, like C-V2X (5G-V2X, LTE-V2X, etc.), are also being widely developed and field tested for the coming era of connected and autonomous vehicles. However, although the versatile connections and information exchange among ICVs, external devices and human beings, provide vehicles better and faster perception of surrounding environments, then more conveniences and better driving experiences for users, they also bring forward a series of intrusion portals for malicious attackers, which threaten the safety of drivers and passengers.

Intelligent and Connected Vehicle is a high-tech complex that concentrates on the application of automobile engineering, artificial intelligence, computer, microelectronics, automatic control, communication technology, big data, edge/cloud computing and other professional knowledge. This book will comprehensively and systematically introduce the advanced technologies and security threats of ICVs from the aspects of automotive technology development, on-board sensors, vehicle networking, automobile communications, intelligent transportation, big data, cloud computing, etc. Then, through some typical automobile attack cases, readers can have deeper understanding of the working principle of ICVs, so that they can test vehicles more objectively and scientifically, find the existence of vulnerabilities and security risks, and take corresponding protective measures to prevent malicious attacks. Technical topics discussed in the book include but not limited to:

- ECU and Vehicular Bus Security;
- Intra-vehicle Communication Security;
- V2X Communication Security;
- VANET Security;
- Unmanned Driving Security and Navigation Deception.

List of Figures

Figure 1.1 The topology of LIN bus. 4
Figure 1.2 LIN standards information. 5
Figure 1.3 The topology of CAN bus. 6
Figure 1.4 The data frame structure of FlexRay. 9
Figure 1.5 The corresponding relationship among OSI model, MOST node model, and MOST network model. . . 10
Figure 1.6 The data frame structure of MOST50. 11
Figure 1.7 The data frame structure of CAN. 15
Figure 1.8 CAN remote frame. 16
Figure 1.9 CAN output voltages when sending a message. . . 18
Figure 1.10 The receiving model for messages on CAN bus. . . 18
Figure 1.11 The connection model of CAN bus, transceiver, and controller. 19
Figure 1.12 The access process of the CAN bus. 20
Figure 1.13 CAN packet acquisition. 21
Figure 1.14 CANTest data analysis. 22
Figure 1.15 Start CAN-Pick to connect to the car bus transceiver. 23
Figure 1.16 The interface of the fuzz testing tool. 24
Figure 1.17 The pin number of the OBD-II interface. 25
Figure 1.18 The general communication architecture of the OBD box. 26
Figure 1.19 Various brands of OBD-II box. 27
Figure 1.20 Remote update process of OBD box. 32
Figure 1.21 Attack OBD box through malicious server. 33
Figure 1.22 Luxgen U5 SUV. 36
Figure 1.23 OBD-II diagnostic equipment. 37
Figure 1.24 The experimental results of malicious tampering in automobile dashboard. 39
Figure 1.25 The working principle of automobile remote key. . 40
Figure 1.26 Bypass the KCU verification mechanism and illegally control the door lock switch. 41

Figure 1.27 Bypass the BCM verification and control the car lights and wipers from the CAN bus. 42
Figure 1.28 Experimental vehicles: Luxgen U5 SUV and Buick Regal. 43
Figure 1.29 MTL network structure. Illegal driver detection, legal driver identification, and driving behaviour evaluation share the same data source and similar feature space. 45
Figure 1.30 Experimental road map. Each driver makes six circles around a fixed route. 47
Figure 1.31 Accuracy and loss curves of legal driver identification in Luxgen U5 SUV and Bucik Regal. (a) Legal driver identification results in Luxgen U5 SUV. (b) Legal driver identification results in Buick Regal. . 49
Figure 1.32 Accuracy and loss curves of driving behaviour evaluation in Luxgen U5 SUV and Buick Regal. (a) Driving behaviour evaluation results in Luxgen U5 SUV. (b) Driving behaviour evaluation results in Buick Regal. 50
Figure 2.1 In-vehicle infotainment. 56
Figure 2.2 The block diagram of RKE system. 58
Figure 2.3 The block diagram of PKE system. 61
Figure 2.4 An illustration of TPMS. 67
Figure 2.5 Specifications of a TPMS provided by NXP. This figure needs to be removed. 68
Figure 2.6 The block diagram of a real-time eavesdropping system. 71
Figure 2.7 A demonstration of packet spoofing based on the eavesdropping system. 72
Figure 2.8 In-vehicle Wi-Fi. 73
Figure 2.9 An illustration of Wi-Fi attack. 73
Figure 2.10 The encryption process of WEP protocol. 75
Figure 2.11 The encryption and integrity of the WPA2 protocol. 77
Figure 2.12 The relationship of keys and parameters in four-way handshake. 78
Figure 2.13 The four-way handshake process. 79
Figure 2.14 The simultaneous authentication of equals handshake in WPA3. 81
Figure 2.15 The Bluetooth legacy authentication. 87

Figure 2.16 The vehicle radio system. 92
Figure 2.17 The working principle of an AM radio. 92
Figure 2.18 The working principle of an FM radio. 93
Figure 2.19 An illustration of FM attack process. 95
Figure 2.20 The vehicle-mounted GPS navigation system. . . . 97
Figure 2.21 Spoofing a vehicle to introduce an erroneous result for position. 100
Figure 2.22 A demonstration of the attack experiment against RKE system using hackRF One. 102
Figure 2.23 The radio signal from a key fob recorded by hackRF One using GNU radio. 102
Figure 2.24 The radio signal of replay attack. 103
Figure 2.25 The process of TSP attack. 104
Figure 2.26 The result of infiltrating TSP. Privacy information and current operating status of the vehicle are obtained from the malicious scripts: (a) the status of the car officially shown; (b) privacy exposed by the attack scripts. 106
Figure 2.27 The position information of the car which is sent to the user's phone through the TSP: (a) the current car's location officially shown; (b) real-time tracking realized by malicious scripts. 107
Figure 2.28 A typical TPMS architecture with four antennas [8]. 108
Figure 2.29 The comparison of FFT and signal strength time series between TSP-A and TSP-B sensors [8]. . . . 109
Figure 2.30 Dash panel snapshots: (a) the tire pressure of left front tire displayed as 0 psi and the low tire pressure warning light was illuminated immediately after sending spoofed alert packets with 0 psi; (b) the car computer turned on the general warning light around 2 seconds after keeping sending spoofed packets [8]. 110
Figure 2.31 Two goals of our threat model. (a) Modify the longitude and latitude of destination before navigation map app plans route to manipulate the victim's destination and route. (b) Change the longitude and latitude of waypoint before navigation map app plans route to manipulate the victim's waypoint and route. 112

Figure 2.32 The detailed steps of navigation map application intrusion. 113
Figure 2.33 The experimental results of Navidog routing falsified by remote spoofing attack. 114
Figure 3.1 Illustration of a representative unmanned vehicle. . 118
Figure 3.2 Three main functional modules in unmanned driving. 119
Figure 3.3 Applications of various sensors in unmanned vehicles. 120
Figure 3.4 Typical structure of vertical control. 124
Figure 3.5 Typical structure of lateral control. 125
Figure 3.6 Data flow in unmanned vehicle's perception, planning, and control systems. 126
Figure 3.7 Ultrasonic radar mounted on the rear of the vehicle. 127
Figure 3.8 Working principle of ultrasonic radar. 130
Figure 3.9 Illustration of jamming attack and spoofing attack. . 132
Figure 3.10 Comparison of transmitting signal and echo signal. 136
Figure 3.11 Doppler shift occurs between the frequency of echo signal and the transmitted signal. 137
Figure 3.12 Target recognition schematic of millimeter wave radar. 138
Figure 3.13 Architecture of camera system. 144
Figure 3.14 Working principle of on-board LiDAR system. . . . 150
Figure 3.15 Experiment settings of attacks against ultrasonic radar. 154
Figure 3.16 Jamming and spoofing attack results on Tesla Model S [11]. 154
Figure 3.17 Experiment settings of attacks against millimeter wave radar. 156
Figure 3.18 Jamming attack results on Tesla Model S [11]. . . . 157
Figure 3.19 Experiment settings of blinding attack against camera sensor. 158
Figure 3.20 Blinding attack against camera sensor with LED spot and laser [11]. 159
Figure 3.21 Experiment settings of relay attack against LiDAR. 161
Figure 3.22 Experiment results of relay attack against LiDAR [7]. 161
Figure 3.23 Attack window of spoofing attack against LiDAR. . 162

Figure 3.24 Experiment settings of spoofing attack against LiDAR. 163
Figure 3.25 The controllable factors of spoofing attack. 164
Figure 3.26 Spoofing LiDAR with one copy and multiple copies [7]. 164
Figure 3.27 LiDAR scanning results of forged wall tracking [7]. 164

List of Tables

Table 1.1 Classification of Automobile Protocols. 2
Table 1.2 Main Threats to Automobile Bus. 13
Table 1.3 Frame Format of CAN Protocol. 14
Table 1.4 Main Security Threats of the OBD Box. 28
Table 1.5 Data Form in CAN Bus. 44
Table 1.6 Confusion Matrix of Illegal Driver Detection in **Luxgen U5 SUV/Buick Regal** (%). 48
Table 1.7 Performance Index and Results of Illegal Driver Detection in **Luxgen U5 SUV/Buick Regal**. 48
Table 1.8 Confusion Matrix of Driving Behaviour Evaluation in **Luxgen U5 SUV/Buick Regal** (%). 50
Table 2.1 The Comparison Among WEP, WPA, WPA2, and WPA3. 83
Table 2.2 Threats and Vulnerabilities of Various Versions of Bluetooth. 88
Table 3.1 Technical Comparison of Main Vehicle Sensors. . . . 121
Table 3.2 Parameter Comparison of Millimeter Wave Radar. . . 134
Table 3.3 Advantages and Disadvantages of Monocular Camera. 142
Table 3.4 Advantages and Disadvantages of Binocular Camera. 142
Table 3.5 Advantages and Disadvantages of Mechanical LiDAR. 148
Table 3.6 Advantages and Disadvantages of Solid-State LiDAR. 148
Table 3.7 Advantages and Disadvantages of Hybrid Solid-State LiDAR. 149

List of Abbreviations

ACAS	automatic collision avoidance system
ACC	adaptive cruise control
ACK	acknowledgement
ACL	asynchronous connection-less
ADAS	advanced driving assistance system
AEB	autonomous emergency braking
AES	advanced encryption standard
AES-CCM	advanced encryption standard-counter mode with cipher block chaining message authentication code
AM	amplitude modulation
AP	access point
AP	automatic parking
APA	auto parking assist
API	application programming interface
ARP	address resolution protocol
ARQ	automatic replay request
ASK	amplitude shift keying
BMW	bavarian motor work
CAN	controller area network
CCD	charge-coupled device
CCMP	cipher block chaining message authentication code protocol
CFCW	constant frequency continuous wave
CMOS	complementary metal oxide semiconductor
CRC	cyclic redundancy check
CSRK	connection signature resolving key
DARPA	defense advanced research projects agency
DBC	database CAN
DDoS	distributed denial of service
DL	deep learning
DLC	data length code

DMS	driver monitor status
DOS	denial of service
EAP	extensible authentication protocol
ECDH	elliptic curve Diffie-Hellman
ECDSA	elliptic curve digital signature algorithm
ECUs	electronic control units
EMI	electro-magnetic interference
FCW	forward collision warning
FEC	forward error correction
FFT	fast Fourier transform
FIPS	federal information processing standard
FM	frequency modulation
FMCW	frequency-modulated continuous wave
FSK	frequency shift keying
GNSS	global navigation satellite system
GPS	global positioning system
ICV	integrity check value
ICVs	intelligent and connected vehicles
IDE	identifier extension
IF	intermediate frequency
IoV	internet of vehicles
IRK	identity resolving key
ISM	industrial scientific medical
ISO	international standardization organization
IV	initialization vector
IVI	in-vehicle infotainment
KCK	key confirmation key
K-L	Karhunen-Loeve
LAN	local area network
LDW	lane departure warning
LE	low-energy
LF	low-frequency
LFSR	linear feedback shift register
LiDAR	light detection and ranging
LIN	local interconnect network
LKA	lane keeping assist
LNA	low noise amplifier
LSTM	long short term memory
LTK	long-term key

MAC	media access control
MCS	master control station
MCU	micro control unit
MEMS	micro-electro-mechanical system
MIC	message integrity code
MITM	man-in-the-middle
MOST	media oriented systems transport
MTL	multi-task learning
NASA	national aeronautics and space administration
NAT	network address translation
NHTSA	national highway traffic safety administration
OBD	on-board diagnostics
OPA	optical phased array
OSI	open system interconnection
OTA	over the air
PDS	pedestrian detection system
PI	planned identification
PID	proportion integral differential
PIN	personal identification number
PKE	passive keyless entry
PMK	pairwise master key
PMMA	polymethyl methacrylate
PRSG	pseudo-random sequence generator
PS	project service
PSK	phase shift keying
PSK	pre-shared key
PTK	pairwise transient key
PVT	position/speed/time
RADIUS	remote authentication dial in user service
RCE	remote code execution
RDS	radio data broadcasting standard
RF	radio frequency
RFID	radio frequency identification
RKE	remote keyless entry
RNG	random number generator
RT	RadioText
RTR	remote transmission request
SAE	simultaneous authentication of equals
SAE	society of automotive engineers

SCI	serial communications interface
SCO	synchronous connection oriented link
SCP	secure copy protocol
SDR	software defined radio
SLAM	simultaneous localization and mapping
SOF	start of frame
SRR	substitute remote request
SSP	secure simple pairing
STA	station
SVC	surround view cameras
SVDD	support vector domain description
TCU	transmission control unit
TK	temporal key
TKIP	temporal key integrity protocol
TPMS	tire pressure monitor system
TPS	tire pressure sensor
TSP	telematics service provider
TSR	traffic sign recognition
UART	universal asynchronous receiver/transmitter
UBIFS	unsorted block image file system
UHF	ultra high frequency
UPA	ultra-sonic parking assist
USRP	universal software radio peripheral
UUID	universally unique identifier
V2X	vehicle-to-everything
WEP	wired equivalent privacy
WLAN	wireless local area network
WPA	Wi-Fi protected access

1

Vehicle Bus Security

1.1 Vehicle Bus

1.1.1 Overview of Vehicle Bus

With the control of automotive systems gradually changing to automation and intelligence, the vehicle electrical network has become increasingly complex. Most of the traditional vehicle electrical networks adopt a point-to-point single communication mode, with few connections between each other, which will inevitably form a huge wiring system. According to statistics, in a high-grade car with traditional wiring methods, the wire length can reach 2000 m, and the number of electrical nodes can reach 1500, which will be doubled every 10 years. This further exacerbates the contradiction between the bulky wiring harness and the limited available space on the car. Regardless of the aspects of material cost or work efficiency, the traditional wiring method cannot adapt to the development of modern automobile. In addition, in order to meet the real-time requirements of various electronic systems, it is necessary to share the public data (such as engine speed, wheel speed, throttle pedal position, and other information), while the real-time requirements of each control unit are different. Therefore, the traditional electrical network has evolved and the new vehicle bus technology has emerged.

As a special internal communication network, vehicle bus connects the internal electronic parts, which is widely used in automobiles, buses, trains, industrial and agricultural vehicles, and so on. In order to meet some special requirements of automobile bus control (such as real-time data transmission and high noise immunity), some professional automobile bus protocols have been developed. These protocols include controller area network (CAN), local interconnect network (LIN), FlexRay, media oriented systems transport (MOST), and others.

1.1.2 Categories of Vehicle Bus

At present, most of the vehicle buses are divided into four classifications: Class A, Class B, Class C, and Class D according to the protocol characteristics by the automotive network committee of the Society of Automotive Engineers (SAE). Table 1.1 illustrates the classifications of various automobile protocols.

Class A: A multiplex wiring system which reduces wiring by transmitting and receiving multiple signals over the same signal bus. The multi-purpose bus replaces individual wires performing the same function. Normally, Class A defines general purpose universal asynchronous receiver/transmitter (UART) communication with bit rates below 10 kbps. LIN specification is the most representative of Class A. It is a new low-cost open serial communication protocol jointly launched by Motorola, Audi, and other enterprises. It is mainly used in distributed electric control system in cars, especially in digital communication for intelligent sensors or actuators.

Class B: A multiplex wiring system which transmits data between nodes. The nodes replace the existing stand-alone modules. Class B is used as a non-critical 10 to 125 kbps bus. CAN is the most typical bus of Class B. In the past, CAN network was only used for data communication between internal measurement and executive parts of automobiles. In 1993, International Standardization Organization (ISO) officially promulgated the international standard (ISO11898-1) of road transportation vehicle – digital information exchange – high-speed communication CAN. In recent years, with the rapid development of intelligent and connected vehicles, CAN is widely used in automobiles.

Class C: A multiplex wiring system which reduces wiring by using real-time high-data signals and is operated between 125 kbps and 1 Mbps. Class C is mainly used in the vehicle power system where the real-time requirements

Table 1.1 Classification of Automobile Protocols.

Category	Bus Name	Communication Rate	Application
A	LIN	Below 10 kbps	Headlights, lights, door locks, power seats, etc.
B	CAN	10 kbps to 125 kbps	Automotive air conditioning, electronic instructions, fault detection, etc.
C	FlexRay	125 kbps to 1 Mbps	Engine control, steering-by-wire system, anti-lock braking system, suspension control, etc.
D	MOST/1394	More than 2 Mbps	Multimedia entertainment system, car navigation system, etc.

of communication are relatively high, and it mainly serves the power transmission system. Nowadays, most automobile manufacturers use "high-speed CAN" as Class C, which is essentially the part of ISO 11898-1 with bit rate higher than 125 kbps. In the United States, special communication protocol SAE J1939 is widely used in trucks, trailers, courses, construction machinery, and agricultural power equipment.

Class D: A high-performance network for multimedia devices and high-speed data stream transmission. The bit rate is generally above 2 Mbps. It is mainly used for CD players and other display devices. Class D has recently been adopted into SAE classification of bus. Its bandwidth range is quite large, and several transmission media are used. It is divided into three categories: low speed (represented by IDB-C), high speed (represented by IDB-M), and wireless (represented by Bluetooth).

1.1.2.1 LIN Bus

LIN is used as an in-vehicle communication and networking serial bus between intelligent sensors and actuators operating. It is mainly used as auxiliary network or sub-network of CAN bus. When the bandwidth requirement is not high, the function is simple, and the real-time requirement is low, such as the auto body electronics (including air conditioning system, doors, seats, switch panel, intelligent wipers, and sunroof actuators). The use of LIN bus can effectively simplify the network harness and improve the efficiency and reliability of network communication.

LIN bus has the following characteristics.

(1) LIN bus has low manufacturing cost and is based on the universal UART/serial communications interface (SCI). Almost all microcontrollers can be equipped with LIN bus.

(2) LIN bus only needs to be equipped with very few signal lines to meet the requirements of ISO9141 standard.

(3) The transmission rate of LIN bus is up to 20 kbps.

(4) The single-master / multi-slave mode of LIN bus does not need arbitration mechanism.

(5) The slave node of LIN bus can realize self-synchronization without crystal oscillator or ceramic oscillator, which saves the hardware cost of slave device.

(6) The delay time of signal transmission in LIN bus is very short and can be ignored.

(7) In LIN bus, the slave node can be added to the network without changing the hardware and software of the slave node.

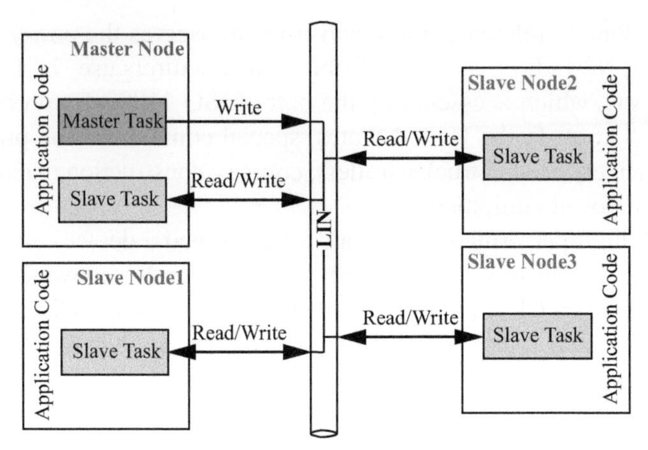

Figure 1.1 The topology of LIN bus.

(8) The number of nodes on a LIN bus is usually less than 12, with a total of 64 identifiers.

The topology of LIN bus is shown in Figure 1.1. LIN bus is based on SCI or UART data format and uses a master–slave approach, having one master node and one or more slave nodes. All nodes contain a subordinate communication task which is decomposed into sending and receiving tasks, and the master node also contains an additional master sending task. The LIN bus does not need to resolve bus collisions because only one message is allowed on the bus at a time.

All communication on LIN bus is initiated by the master task in the master node. The master task determines the current communication content based on the schedule, sends the corresponding frame header, and allocates the frame channel for the message. After receiving the frame header, the slave node on the bus interprets the identifier to determine whether it should respond to the current communication and what kind of response. Based on this message filtering method, LIN bus can realize multiple data transmission modes, and a single message frame can be simultaneously received and utilized by multiple nodes.

LIN standards information is shown in Figure 1.2. The master node sends a message header containing synchronization break, synchronization field, and message identifier. After receiving and filtering the identification code, the slave task is activated and begins the transmission of the message response. The response contains 2, 4, or 8 data bytes and a checksum byte. The header and the response parts form LIN standards information.

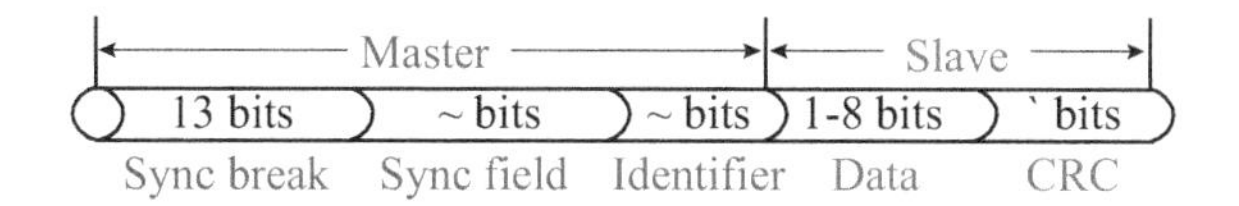

Master, Sync break [13 bits]: used to identify the start of the fram
Master, Sync field [alternate 1-0 sequences]:used by the slave node for clock synchronization
Master, Identifier [6-bit long message ID and a 2-bit long parity field]
Master, Message ID [2, 4 or 8 data bytes]: optional message length information
Slave transmission ~[1-8 data bytes]: data bytes
Slave transmission ~[8 bit]: checksum

Figure 1.2 LIN standards information.

1.1.2.2 CAN Bus

CAN is a serial communication protocol standardized by the ISO. In the automotive industry, a variety of electronic control systems have been developed for safety, comfort, convenience, low power consumption, and low cost. Due to the different data types and reliability requirements used for communication between these systems, there are many cases of multiple buses, and the number of wire harness increases accordingly. In order to meet the needs of "reducing the number of wiring harnesses" and "high-speed communication with large amounts of data through multiple LANs", Bosch, a German electric company, developed a CAN communication protocol for automobiles in 1986. Since then, CAN has been standardized by ISO11898 and ISO115119 and has been the standard protocol of automobile network in Europe.

The high performance and reliability of CAN has been recognized and widely used in industrial automation, ship, medical equipment, industrial equipment, and other aspects. Its appearance provides powerful technical support for distributed control system to realize real-time and reliable data communication between nodes.

Depending on the data transmission speed, CAN buses can be divided into two types: high-speed CAN (ISO 11898-2) and low-speed CAN (ISO 11898-3). The data rate of high-speed CAN is 125 kbps to 1 Mbps, which is applied to nodes with high real-time requirements, such as engine control, anti-lock brakes, transmission control, etc. Low-speed CAN has a data rate of 5–125 kbps, which is mainly used in the field of comfort and entertainment, such as power windows, power seats, lightings, dashboard, etc. These nodes do not require high real-time performance, and the distribution is relatively scattered. The transmission speed of low-speed CAN is well adapted to the

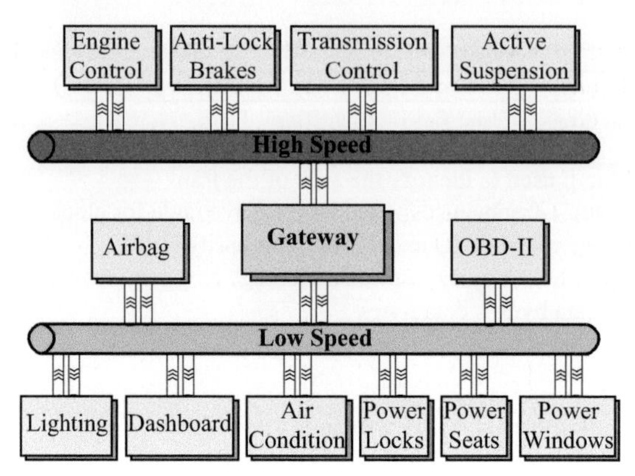

Figure 1.3 The topology of CAN bus.

requirements. Figure 1.3 shows the topology of CAN bus in automobile. CAN bus devices with different speed types cannot be directly connected to the same bus, and they need to be isolated by gateway.

CAN bus is a multi-master bus. The communication medium can be twisted pair, coaxial cable, or optical fibre, and the communication rate is up to 1 Mbps. The features of CAN protocol are as follows:

- CAN bus completes the frame processing of communication data. The physical layer and data link layer functions of CAN protocol are integrated in the CAN bus communication interface, which can complete the frame processing of communication data, including bit filling, data block coding, cyclic redundancy check (CRC), priority discrimination, and so on.
- One of the most important features of CAN protocol is that the traditional station address coding is replaced by the coding of communication data block. The advantage of this method is that the number of nodes in the network is not limited theoretically. The identifier of data block can be composed of 11- or 29-bit binary numbers. Therefore, two or more different data blocks can be defined. This coding method can also make different nodes receive the same data at the same time, which is very useful in distributed control system. The maximum length of the data segment is 8 bytes, which can meet the general requirements of control command, working state, and test data in general industrial field. At the same time, 8 bytes will not occupy the bus for a long time so as to ensure the real-time communication. CAN protocol adopts CRC inspection and provides corresponding error handling function to ensure the reliability

of data communication. Because of its excellent characteristics, high reliability, and unique design, CAN is especially suitable for the interconnection of industrial process monitoring equipment. Therefore, it has been paid increased attention by the industry and has been recognized as one of the most promising field-buses.

- CAN can realize free communication among nodes. CAN bus adopts multi-master competitive bus structure, which has the characteristics of multi-master station operation, decentralized arbitration serial bus, and broadcast communication. Any node on CAN bus can actively send information to other nodes in the network at any time without priority; so free communication can be realized among all nodes. CAN bus protocol has been certified by ISO, and its technology is relatively mature. The control chip has been commercialized with high-cost performance. It is especially suitable for data communication between distributed measurement and control systems.
- The structure of CAN bus is simple, only two wires are connected with the outside, and the error detection and management module is integrated in the CAN bus.
- CAN nodes have no master–slave relationship in data communication. One node can initiate data communication to any other (one or more) nodes, and the communication order is determined by the priority of each node's information. When multiple nodes initiate communication at the same time, high priority nodes send data first, which will not cause communication line congestion.

Although CAN bus is widely used in industrial automation, ship, medical equipment, industrial equipment, and other fields, it also has certain limitations.

- Because of the arbitration characteristic of CAN bus, even if the message is sent periodically to the bus, it cannot guarantee that the node can receive the message periodically. Therefore, CAN is not suitable for time sensitive applications.
- The maximum transmission rate is only 1 Mbps, and the bandwidth of data transmission or video and audio transmission is insufficient for the application of autonomous driving.
- For simple applications, high-cost CAN bus is reliable but wasteful. Compared with CAN, LIN bus has the advantages of low cost and is more suitable for windows, seats, air conditioning, and other equipment.

1.1.2.3 FlexRay Bus

FlexRay bus is a communication standard jointly developed by Bavarian Motor Works (BMW), Philips, Freescale, and Bosch, which is specially designed for in-vehicle network. FlexRay bus combines event triggered and time triggered data sending–receiving methods, which makes it have the characteristics of efficient network utilization and system flexibility. The FlexRay protocol ensures that information delay and jitter are minimized and transmission is as synchronous and predictable as possible, even if the driving environment is harsh and changeable. This is very important for applications that require continuous and high-speed performance, e.g., wire braking, wire steering, etc.

FlexRay has the features of high speed, reliability, and safety. FlexRay physically communicates through two separate buses, each with a data rate of 10 Mbps. FlexRay also offers many reliable features that traditional networks do not have, especially the redundant communication capability of FlexRay, which can completely copy the network configuration through hardware and monitor the progress. FlexRay also provides flexible configuration to support a variety of topologies, such as bus, star, and hybrid topologies. FlexRay does not guarantee the safety performance of the system, but it has a lot of functions, which can support the design of safety-oriented system (such as wire control system).

FlexRay can be applied in passive bus and star network topologies, as well as in their combined ones. Both topologies support dual-channel electronic control units (ECUs), which integrate multiple system-level functions to save production cost and reduce complexity. The dual-channel architecture provides redundancy and doubles the available bandwidth. The maximum data transmission rate of each channel reaches 10 Mbps.

FlexRay bus currently only has protocols at the physical layer and data link layer. Figure 1.4 shows the data frame format of FlexRay, including header segment, payload segment, and trailer segment.

The header segment consists of 5 bytes (40 bits), including the following contents:

- The reserved bit (1 bit) to prepare for future expansion.
- The payload preamble indicator (1 bit) indicates the vector information of the frame's Payload segment. In a static frame, this bit indicates the NWVector, while in a dynamic frame, it indicates the information ID.
- The null frame indicator (1 bit) indicates whether the data frame of load segment is zero.

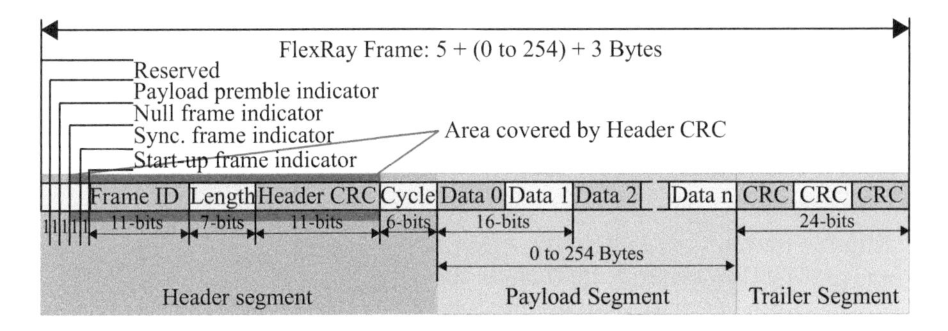

Figure 1.4 The data frame structure of FlexRay.

- The sync frame indicator (1 bit) indicates that this is a synchronous frame.
- The start-up frame indicator (1 bit) indicates whether the frame that the node sends is the start frame.
- The frame ID (11 bits) indicates the ID assigned to each node in the system design process (valid range: 1–2047).
- Length (7 bits), indicating the data length of the load section.
- The header CRC (11 bit) indicates the CRC calculation value of synchronous frame indicator and start frame indicator, as well as the frame ID and frame length period calculated by the host.

The payload segment consists of three parts:

- The Data, 0–254 bytes.
- Information ID, defined by the first two bytes of the payload segment, can be used as filterable data at the receiver.
- NWVector, which must be between 0 and 10 bytes in length and the same as all nodes.

The trailer segment of the frame only contains a single data field, namely the CRC, including header CRC and data CRC. These CRC values will change seed values on the connected channel to prevent incorrect correction.

1.1.2.4 MOST Bus

MOST bus is a vehicle bus jointly completed by BMW, DaimlerChrysler, Harman/Becker, and Oasis Silicon Systems. In 1998, the participants set up an independent entity, MOST company, which controls the definition of bus.

MOST bus is designed to meet the requirements of the vehicle environment. This new fibre-based network supports a data rate of 24.8 Mbps and has the advantages of reducing weight and electro-magnetic interference (EMI) compared with other copper-based networks. Different from CAN bus, MOST bus is not connected by twisted pair but by a single optical fibre and transmitted by optical pulse. In the physical layer, the MOST bus transmission medium is polymethyl methacrylate (PMMA) optical fibre with plastic sheath and inner core of 1 mm. Suppliers can bundle a bunch of optical fibres into optical cables like wires. Optical fibre transmission uses 650 nm (red) LED transmitter (650 nm is the low-loss "window" in PMMA spectral response). Data is sent in a 50-Mbaud bi-phase encoding manner with a maximum data rate of 24.8 Mbps.

The MOST bus specifications define the physical layer (electrical and optical parameters) as well as the application layer, network layer, and medium access control layer. MOST node model can be divided into photoelectric transceiver, network interface controller, basic network service, high-level network service, etc. MOST network structure corresponds to the seven-layer network model of ISO. Figure 1.5 shows the corresponding relationship among open system interconnection (OSI) model, MOST node model, and MOST network model.

MOST allows for a variety of topologies, including star and ring, and most of automotive devices use ring layouts. A MOST network can have 64 nodes at most. Once the car is powered on, all the MOST nodes in the network will be activated. The default state of the MOST node is through when it is

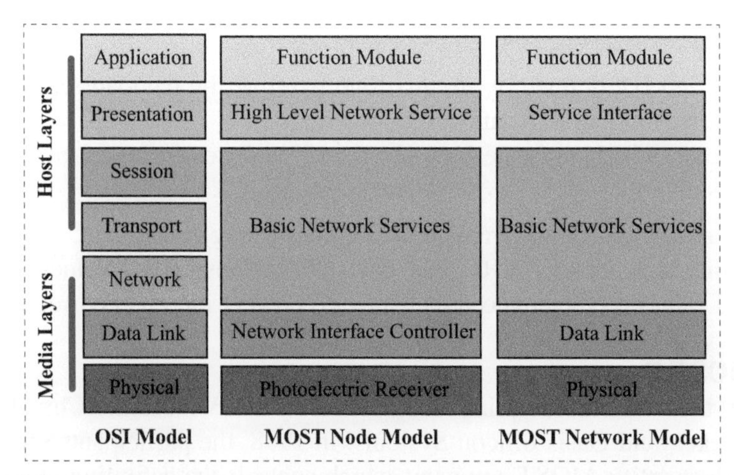

Figure 1.5 The corresponding relationship among OSI model, MOST node model, and MOST network model.

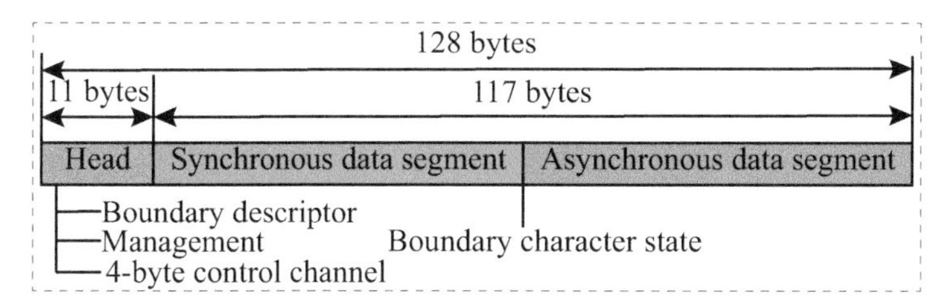

Figure 1.6 The data frame structure of MOST50.

powered on, that is, the incoming data is transmitted directly from the receiver to the transmitter to keep the loop unblocked. In MOST50 network, the basic unit of data transmission is data frame, and the length of each data frame is 128 bytes. Figure 1.6 shows the structure of the MOST50 data frame.

MOST network supports three different types of data transmission: control data transmission, synchronous data transmission, and asynchronous data transmission. Since only 4-byte control data is transmitted in each data frame, multiple data frames are needed to transmit one control message. Synchronous data is real-time data, such as audio and video data. Asynchronous data is non-real-time large data, such as Internet data, GPS map data, etc. Asynchronous data is transmitted in the network aperiodically by token ring. The receiver ensures the accuracy of data transmission by CRC.

1.1.2.5 Automotive Ethernet Bus

In recent years, the insufficient bandwidth of CAN bus has been troubling automotive electronic engineers. Once the configuration of the original vehicle is improved and its functions are increased, problems may arise due to too many nodes on the bus and too high load rate, thus affecting the ability of sending and receiving messages on the network of the vehicle. If additional automobile gateway is added, the R & D cycle will be extended and the additional cost will be increased. Most importantly, CAN is the current mainstream vehicle network technology. Ethernet has been very mature in our life, industry, and other fields. Its characteristics of huge bandwidth, low cost, and good electrical isolation bring broad prospects for automobile enterprises. Ethernet is the most popular local area network (LAN) architecture at present. It has the obvious advantages of openness, low cost, and widely used software and hardware support.

On-board Ethernet is a physical network used to connect various electrical equipment in a car. Vehicle-mounted Ethernet is designed to meet some special requirements in vehicle environment. For example, the requirements for electrical characteristics of on-board equipment; the requirements for high bandwidth, low latency, and audio/video synchronization of on-board equipment; the demand for network management of vehicle-mounted system. In other words, based on the civil Ethernet protocol, on-board Ethernet has changed the electrical characteristics of the physical interface, and some new standards have been customized according to the requirements of vehicle-borne network. IEEE 802.1 and IEEE 802.3 standards have also been supplemented and revised accordingly by IEEE organizations for on-board Ethernet standards.

Automotive Ethernet enables full duplex operation, sending and receiving data at the same time without waiting. Its packet switching function can realize multiple switching of different devices under various conditions. On-board Ethernet data is address-based message transmission. Each Ethernet message has a source address and a destination address. The switch sends the message to the target receiver according to the destination address, and the receiver can reply according to the original address read from the message.

Currently, the four most important basic topologies of Ethernet are point-to-point, bus, ring, and star topologies. Compared with other traditional in-vehicle networks, Ethernet has great advantages in high bandwidth, low latency, and small cost, making it a promising new in-vehicle network in the future.

1.1.3 Threat Analysis of Vehicle Bus

With the gradual application of the intelligent auxiliary systems in the automobile, such as advanced driving assistance system (ADAS) and automatic collision avoidance system (ACAS), the external network can access the crucial components of the car, such as engines, brakes, airbag, etc. Control information is transmitted through the bus system between components. The bus system basically has no defence against malicious attacks. In particular, the MOST bus is responsible for multimedia transmission, the LIN bus is responsible for door lock switch, and the CAN bus is responsible for multi-master node communication, their communication with wireless network interface is becoming more and more frequent, but the communication data is all unencrypted. In addition, most of the information coding and communication protocols in automobile bus are public. The controller also does not have the corresponding detection procedure to verify whether the arrival information is legal control information.

Table 1.2 Main Threats to Automobile Bus.

Bus Type	Risk	Main Threat
CAN bus	High	Eavesdrop, forge, or tamper with data frames; Interrupt the transmission of high priority normal frames; Forge fault signal frames to deceive the terminal.
LIN bus	Low	A single node is invalidated by the dependency of the master/slave nodes; The LIN bus cannot work normally because of the synchronization feature.
MOST bus	Low	Interfere the signal synchronization via the features of time source; Forge the request information of channels and consume the bandwidth.
FlexRay bus	High	Eavesdrop, forge, or tamper with data frames; Interrupt the transmission of high priority normal frames; Attackers make the controller sleep by creating sleep signal frame.
Ethernet bus	High	distributed denial of service (DDoS) attack; replay attacks; Eavesdrop, forge, and tamper with packets; fuzz testing attacks.

In theory, any controller of the CAN bus, the MOST bus, and the LIN bus can send instructions to any other controller. Therefore, any controller suffering the bus attack will pose a substantial threat to the vehicle communication network. For example, the multimedia buses are connected with the external interface and the Internet, which may make malicious software, such as remote trojan, malicious viruses, intrude into the core system of the vehicle through USB, e-mail, and other means. However, part of the gateways in the vehicle provides simple firewall mechanism, but the MOST bus and the CAN bus also provide powerful and undefended diagnostic interfaces, which enables attackers to break through the vehicle network easily. Table 1.2 lists the main threats to the automobile bus.

At present, the protection of automobile bus system is still stuck at the stage where the source address and destination address of the information received by the controller is verified, and the channel used to transmit the information and the gateway firewall are encrypted. These measures cannot avoid that low-risk networks send information to the high-risk ones. The protective measure is only to close all interfaces when the car starts normally, which would greatly reduce the drivers' experience.

1.2 CAN Bus Vulnerability and Analysis

1.2.1 CAN Bus Architecture

CAN has two kinds of message frames, essentially different in the length of the ID. The format of the CAN2.0A message frame is the standard format

Table 1.3 Frame Format of CAN Protocol.

Frame Type	Frame Application
Data frame	Transmit data from a transmitting unit to a receiving unit
Remote frame	A frame used by a receiving unit to request data from a sending unit that has the same ID
Error frame	Notify other units of errors when errors are detected
Overload frame	Tell that the receiving units are not ready to receive

of the CAN message frame, which has 11-bit identifiers. Networks based on CAN2.0A can only receive messages in this format. The message frame format of CAN2.0B, also called extended message format, has 29-bit identifiers, in which the first 11-bit identifiers are the same as that of CAN2.0A message frame and the last 18-bit identifiers are dedicated to marking the CAN2.0B message frame.

CAN message frames can be mainly divided into four different types according to their applications, as shown in Table 1.3. The data frame is used for transmitting data from the sending unit to the receiving unit. The remote frame is used for requesting to send data. The error frame identifies detected errors, and the overload frame delays the transmission of the next message frame. The detailed description of CAN bus architecture is as follows.

1. Data Frame

The CAN data frame is composed of seven different bit fields, namely, start of frame (SOF), arbitration field, control field, data field, CRC field, acknowledgement field, and end of frame, where the length of the data field can be zero. The following is a brief analysis of the functions of these fields. Figure 1.7 shows the composition of the data frame.

(1) SOF: Mark the start of the data frame or the remote frame that consist of a single "dominant" bit. Nodes are allowed to send signals only when the bus is idle.

(2) Arbitration field: The arbitration field is composed of 11-bit identifiers and remote transmission request (RTR) bit in the normal format, while it is composed of 29-bit identifiers, SRR bit, the other identifier and RTR bit in extensible format.

• RTR: The RTR bit must be dominant in the data and recessive in the remote frame.

• Substitute remote request (SRR): It is always recessive in the extended format.

• Identifier extension (IDE): The IDE bit belongs to the arbitration field for the extended format and the control field for the standard format. While

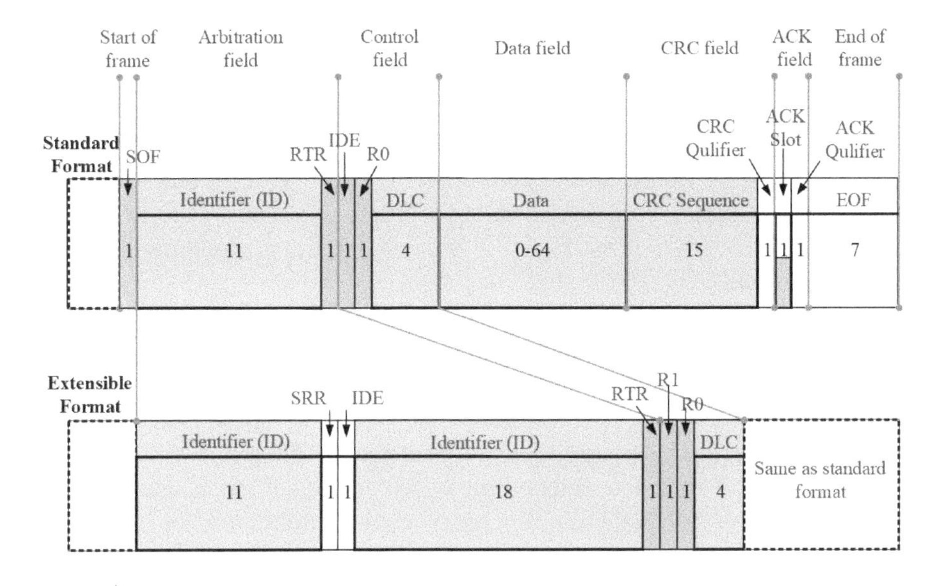

Figure 1.7 The data frame structure of CAN.

the IDE bit is dominant in the standard format, it is recessive in the extended format.

(3) Control field: It consists of six bits, including the data length code (DLC) and two reserved bits for extension in future. A message frame includes DLC, IDE bit of sending dominant level, and reserved bit r0 in the standard format, where DLC and two reserved bits, r1 and r0, must send dominant levels.

(4) Data field: The data area is composed of the transmitted data in the data frame. It can be between 0 and 8 bytes; each byte contains 8 bits, and the highest effective bit is sent first.

(5) CRC field: It includes CRC sequence and CRC delimiter.

(6) Acknowledgement (ACK) field: There are 2 bits in length, including ACK slot and ACK delimiter. In the ACK field, the sending node sends two recessive bits. A receiver that correctly receives a valid message will report this information to the transmitter by transmitting a dominant bit during the ACK slot. All stations that have received a matching CRC sequence report by writing the dominant bit in the ACK slot into the recessive bit of the transmitter. The ACK delimiter is the second bit of the ACK field and must be recessive.

(7) End of frame: Each data frame or remote frame can be defined by a flag sequence, which is composed of seven recessive bits.

2. Remote frame

The node that receives the data can request the source node to send data by sending a remote frame. It consists of six fields: SOF, arbitration field, control field, CRC field, ACK field, and end of frame. It has no data field, and the RTR bit is a recessive level. Figure 1.8 shows the frame structure of the CAN remote frame.

3. Error frame

The error frame is composed of two fields, namely the error flag and the error delimiter. When the receiving node finds that there is an error in the message on the bus, it will automatically send the active error flag, which is six consecutive dominant bits. After detecting the active error flag, other nodes send an error acceptance flag, which consists of six consecutive recessive bits. Since the time when each receiving node finds the error may be different, the actual error flag on the bus may be composed of 6–12 dominant bits. The error delimiter consists of eight recessive bits. When the error flag occurs, each CAN node monitors the bus until it detects a dominant level jump. It means that all nodes have finished sending the error flags at this time and start to send the delimiter of eight recessive levels.

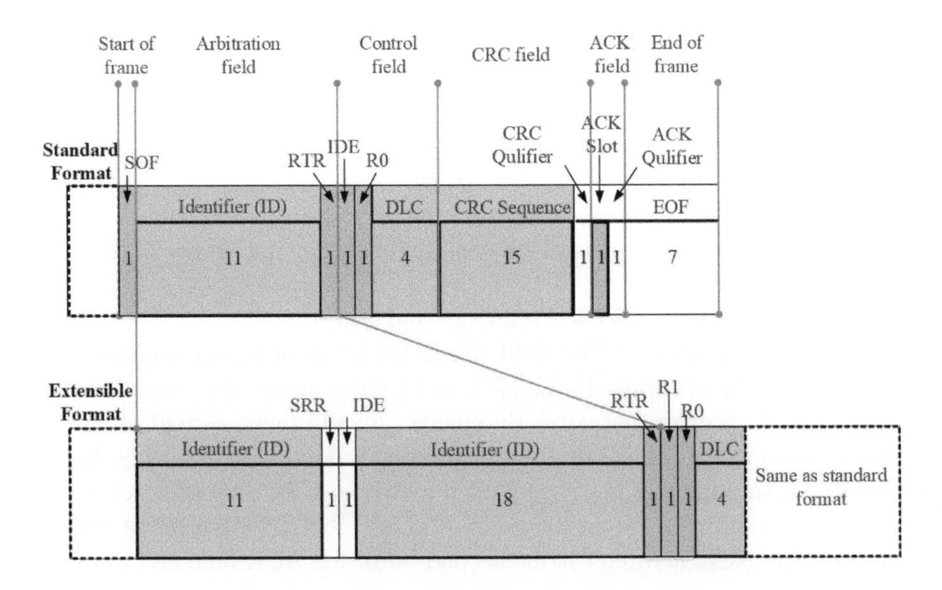

Figure 1.8 CAN remote frame.

4. Overload frame

An overload frame includes two fields, the overload flag and the overload delimiter. There are two types of overload conditions that cause the transmission of the overload flag: one is the internal condition of the receiver that requires the delay of the next data frame or remote frame, and the other is the detection of dominant bits on the first and second bits of the intermittent field. The overload flag is composed of six dominant bits, and the overload delimiter is composed of eight consecutive recessive bits.

After the CAN bus is applied to automobiles, the information security problem has persisted. Although the CAN network unifies the standard protocol in the physical layer and logical link layer, different manufacturers use different data communication protocols in specific data request instructions and response mechanisms. These protocols are secret and incompatible; so CAN bus can effectively prevent attackers from invading.

With the development of the automobile intelligence, more and more products about the Internet of vehicles appear, which makes the internal network communication protocol no longer mysterious. More and more ECUs in cars are connected to the Internet, which makes it easier for malicious attackers to attack the vehicle network. In the existing vehicle attack methods, whether through the OBD-II interface or through the remote wireless connection, it is ultimately necessary to send forged instructions to the ECU through the CAN bus to achieve the purpose of controlling the vehicle.

1.2.2 On-Board CAN Bus Access

Before invading the car, it is necessary to find the CAN bus of the car at first. The CAN bus interface of the car exists in the door, headlight, trunk, in-vehicle infotainment (IVI), and so on. There are many wiring harnesses at these interfaces of the car, among which the CAN bus could be found.

First, the CAN bus has obvious characteristics. It is a twisted pair. After finding it, a multimeter can be used to measure the voltage on both ends of the line. If the voltages of the two wires are about 3.5 and 1.5 V, respectively, they are likely to be CAN bus. When the CAN bus is currently in a sleep state, there is no signal data and no differential voltage. Therefore, if the measured voltage is about 2.5 V, it may also be a CAN bus. In addition, the twisted pair, of which the impedance between two lines is about 120 Ω, is very likely to be the CAN bus. Combined with the above information, it is easy to determine which line is the CAN bus. Figure 1.9 shows the structure of CAN bus twisted pair.

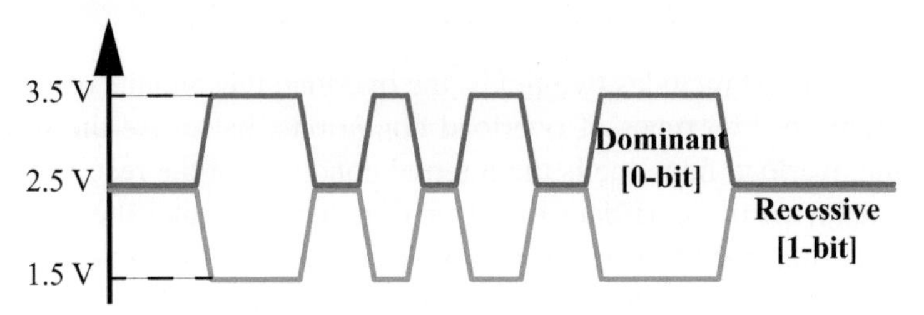

Figure 1.9 CAN output voltages when sending a message.

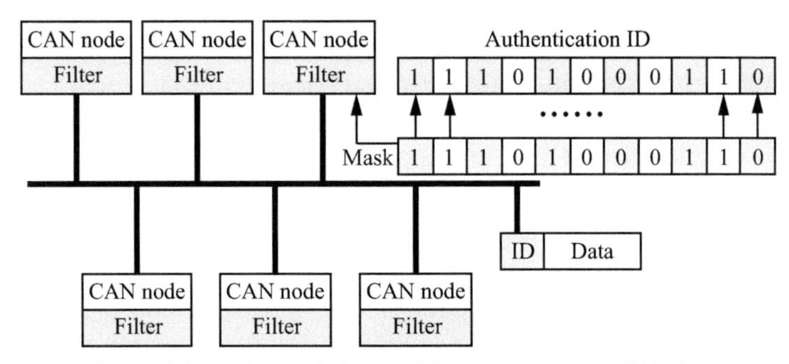

Figure 1.10 The receiving model for messages on CAN bus.

The nodes in the CAN bus are divided into sending nodes and receiving nodes. In the CAN network, when the bus is idle, the node that sends a message to the CAN network becomes the sending node. When the node does not act as the sending node, it serves as the receiving node. In this way, multi-cast or broadcast can be achieved. The sending node in the CAN bus uses the identifier ID to specify the destination address of the transmission of the message, and the receiver filters the received messages according to its own needs so that the nodes in the CAN bus can work in the one-to-one or one-to-many way. Figure 1.10 shows the model of message receiving in the CAN network.

In the CAN bus network, whenever the sending node sends out a CAN message, all nodes connected to this CAN bus filter the message according to the configured node ID and mask and receive data according to their own needs. In the hierarchical architecture of the CAN network, the data link layer and the physical layer are indispensable parts to ensure communication quality, and also the most complicated part of the CAN network. The CAN controller is used to realize the data link layer of the CAN communication protocol, and the CAN module in the general microprocessor represents the

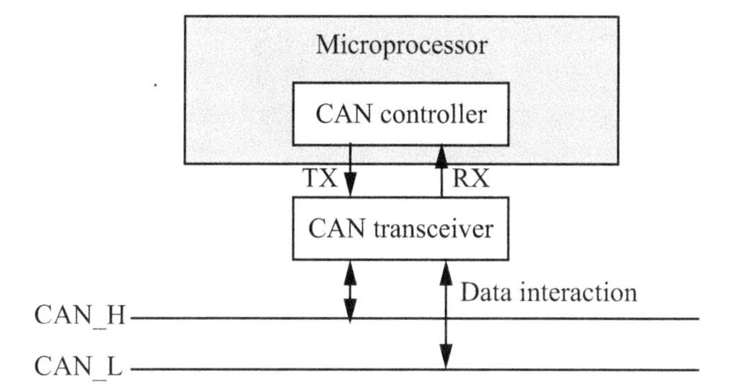

Figure 1.11 The connection model of CAN bus, transceiver, and controller.

CAN controller. The CAN controller is implemented by using a combination of integrated logic circuits inside the microprocessor chip. It is to complete the filling of the CAN frame and convert it into a binary code stream or analyse the received binary code stream.

While the CAN transceiver is in the physical layer of the CAN bus, it is to convert the binary code into the differential signal on the CAN bus to send. A CAN transceiver usually provides two pins, CAN_H and CAN_L, and the signal appears in the form of a differential voltage between the two pins. The two complementary logic values of "dominant" and "recessive" are used on the bus to represent "0" and "1", respectively. Values on the bus are recessive only when the two pins send recessive signals. Figure 1.11 shows the connection model of CAN bus, CAN transceiver, and CAN controller.

In practice, using a multimeter to measure, the high voltage is CAN-H, while the low voltage is CAN-L. After distinguishing the high and low lines, one end of the transceiver is connected with CAN-H and CAN-L, respectively, and the other end is connected with PC to receive/send CAN data. Figure 1.12 shows the access process of the CAN bus, while the PC corresponds to the CAN controller in Figure 1.11.

1.2.3 Reverse CAN Bus Communication

In the experiment, CAN data packets can be obtained via CANalyst-II (an automotive diagnostic tool). The data packets being sent on the CAN bus can be observed through the CANTest software in a real-time manner. Then, researchers need to reverse analyse a large number of CAN packets to find the CAN ID corresponding to the car control command. It is the most basic and

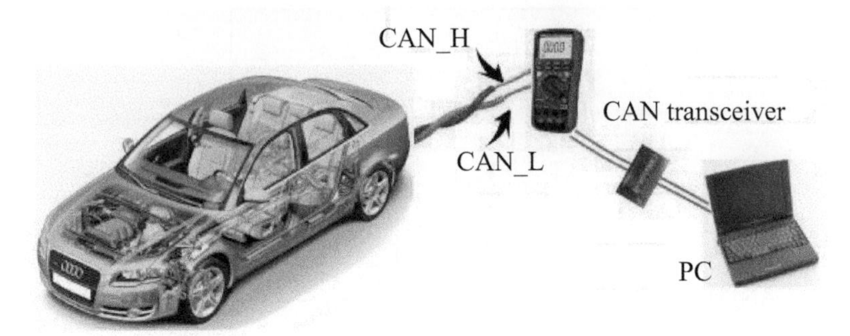

Figure 1.12 The access process of the CAN bus.

important step to reversely analyse the meaning of such CAN data packets. The CANTest interface is shown in Figure 1.13. Note that since the message information collected is from the actual CAN bus, parts of the message data are hidden, the same below.

After reversing the instruction information of these car-controlled data packets, the meaning and principle of work about these data packets are analysed, and feasible attack strategies based on the working principles are formulated. For example, malicious attackers only need to replay a single command to control the car door switch, but they need to combine multiple data packets of CAN ID to control the vehicle speed. To construct such hybrid CAN ID packets, the attacker needs to set a fixed sending interval to send it to the CAN network. The reason for setting the sending interval is to bypass the time detection mechanism of ECU.

Some CAN ID data packets contain a counter, called heartbeat packet, which must be added during an attack to bypass the system check. When the attacker writes the script program, he needs to simulate the change rule of CAN ID data packet bits and send such heartbeat packet data to CAN bus.

In CAN packets, each byte of data has a unique meaning, and it relies on some simple decryption algorithms to calculate it. To crack this kind of data packet, the attacker needs to collect a large number of data packets, deduces the formula reversely, and changes the desired value to control the display of automobile dashboard.

In addition, the idea of fuzzy testing can be used to write a fuzzy testing script tool. Attackers need to constantly send data packets to CAN bus, observe the response of the car, and then judge the function command corresponding to the sent data packets.

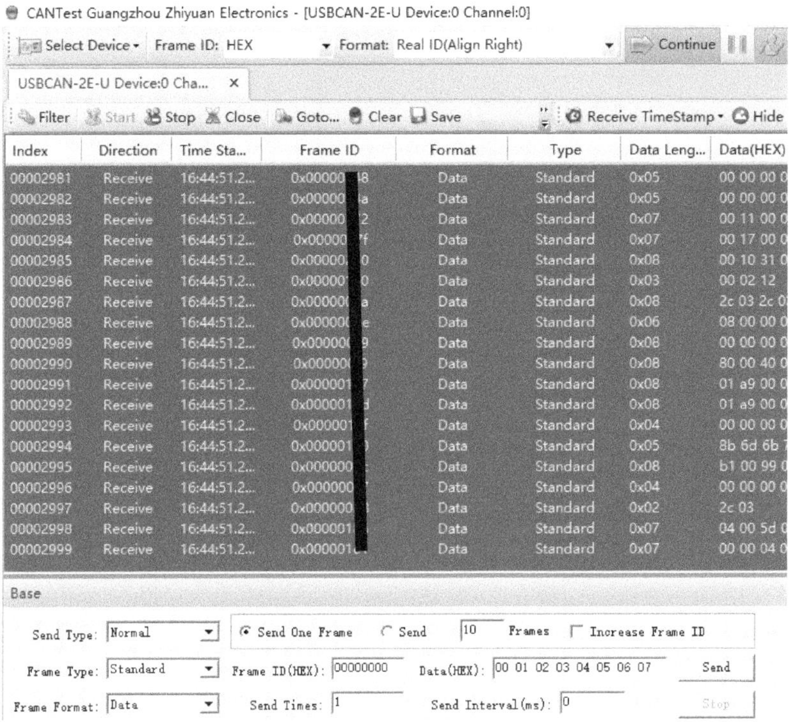

Figure 1.13 CAN packet acquisition.

1.2.4 Analysis of CAN Message Data

The attacker uses the DBC (database CAN) function of CANTest to analyse the control instructions in CAN bus. Among them, the DBC shows all CAN IDs in the current bus, as shown in Figure 1.14. After the car issues an action order, the CAN ID message would change. DBC will mark the changed part in red. By observing which CAN ID changes, which is usually only in an instant, the CAN ID can be determined corresponding to the vehicle control instruction.

By changing the switch state of the door, the DBC is used to observe the door data. When the door changes the switch state, observing which CAN ID has changed can determine whether it is related to the door state. The test results show that when the car does not start and all the electrical equipment in the car remain in the original state, only the state of the door is changed, then the data of some frames on the DBC interface has changed accordingly, as shown in the rectangular box of Figure 1.14.

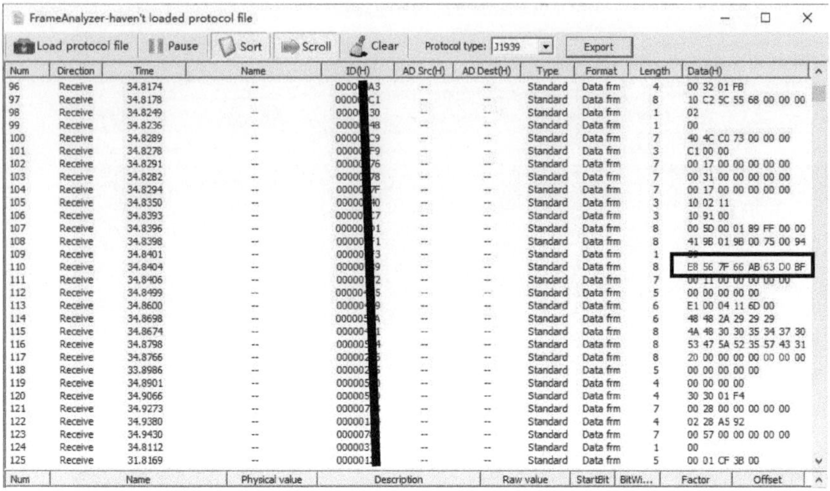

FrameAnalyzer-haven't loaded protocol file

Load protocol file | Pause | Sort | Scroll | Clear | Protocol type: J1939 | Export

Num	Direction	Time	Name	ID(H)	AD Src(H)	AD Dest(H)	Type	Format	Length	Data(H)
96	Receive	34.8174	--	0000[illegible]A3	--	--	Standard	Data frm	4	00 32 01 FB
97	Receive	34.8178	--	0000[illegible]C1	--	--	Standard	Data frm	8	10 C2 5C 55 68 00 00 00
98	Receive	34.8249	--	0000[illegible]30	--	--	Standard	Data frm	1	02
99	Receive	34.8236	--	0000[illegible]48	--	--	Standard	Data frm	1	00
100	Receive	34.8289	--	0000[illegible]C9	--	--	Standard	Data frm	7	40 4C C0 73 00 00 00
101	Receive	34.8278	--	0000[illegible]F9	--	--	Standard	Data frm	3	C1 00 00
102	Receive	34.8291	--	0000[illegible]76	--	--	Standard	Data frm	7	00 17 00 00 00 00 00
103	Receive	34.8282	--	0000[illegible]78	--	--	Standard	Data frm	7	00 31 00 00 00 00 00
104	Receive	34.8294	--	0000[illegible]7F	--	--	Standard	Data frm	7	00 17 00 00 00 00 00
105	Receive	34.8350	--	0000[illegible]40	--	--	Standard	Data frm	3	10 02 11
106	Receive	34.8393	--	00000[illegible]C7	--	--	Standard	Data frm	3	10 91 00
107	Receive	34.8396	--	00000[illegible]1	--	--	Standard	Data frm	8	00 5D 00 01 89 FF 00 00
108	Receive	34.8398	--	00000[illegible]1	--	--	Standard	Data frm	8	41 9B 01 9B 00 75 00 94
109	Receive	34.8401	--	00000[illegible]3	--	--	Standard	Data frm	1	[illegible]
110	Receive	34.8404	--	00000[illegible]9	--	--	Standard	Data frm	8	E8 56 7F 66 AB 63 D0 BF
111	Receive	34.8406	--	00000[illegible]2	--	--	Standard	Data frm	7	00 11 00 00 00 00 00
112	Receive	34.8499	--	00000[illegible]5	--	--	Standard	Data frm	5	00 00 00 00 00
113	Receive	34.8600	--	00000[illegible]9	--	--	Standard	Data frm	6	E1 00 04 11 6D 00
114	Receive	34.8698	--	000005[illegible]A	--	--	Standard	Data frm	6	48 48 2A 29 29 29
115	Receive	34.8674	--	00000[illegible]1	--	--	Standard	Data frm	8	4A 48 30 30 35 34 37 30
116	Receive	34.8798	--	000005[illegible]4	--	--	Standard	Data frm	8	53 47 5A 52 35 57 43 31
117	Receive	34.8766	--	000002[illegible]5	--	--	Standard	Data frm	8	20 00 00 00 00 00 00 00
118	Receive	33.8986	--	000002[illegible]5	--	--	Standard	Data frm	5	00 00 00 00 00
119	Receive	34.8901	--	000005[illegible]	--	--	Standard	Data frm	4	00 00 00 00
120	Receive	34.9066	--	000005[illegible]	--	--	Standard	Data frm	4	30 30 01 F4
121	Receive	34.9273	--	000007[illegible]	--	--	Standard	Data frm	7	00 28 00 00 00 00 00
122	Receive	34.9380	--	000001[illegible]	--	--	Standard	Data frm	4	02 28 A5 92
123	Receive	34.9430	--	000007[illegible]	--	--	Standard	Data frm	7	00 57 00 00 00 00 00
124	Receive	34.8112	--	000003[illegible]	--	--	Standard	Data frm	1	00
125	Receive	31.8169	--	000001[illegible]	--	--	Standard	Data frm	5	00 01 CF 3B 00

Num	Name	Physical value	Description	Raw value	StartBit	BitWi...	Factor	Offset

Figure 1.14 CANTest data analysis.

DBC will list all the current CAN IDs on the vehicle bus. When the state of the car changes, the data bit changed will be marked in red. Therefore, it can control the relevant variables of the car, repeatedly open and close the door, and the switching control instructions of the car door can be analysed by DBC.

After determining which CAN ID is related to the door status, the attackers need to determine which bit controls the status display. The valid information of the CAN message includes different CAN signals, and each CAN signal carries data for options or configurations of a specific function. For example, the data field of a CAN message may contain the status information of the anti-lock brake system, as well as different signals of configuration data and sensor values. The classifier generates an event class for each signal value of the message. That is to say, a CAN ID can have multiple functions, and its data information contains state information and control information of a variety of different functions. Researchers find the CAN ID of the control command, fuzzy test its data bits, and find the bits related to the control command by changing the numbers of the data bits.

1.2.5 Fuzz Testing of CAN Bus

Based on the idea of fuzzy testing, vulnerabilities can be found by providing unexpected input to the target system and monitoring abnormal results. The

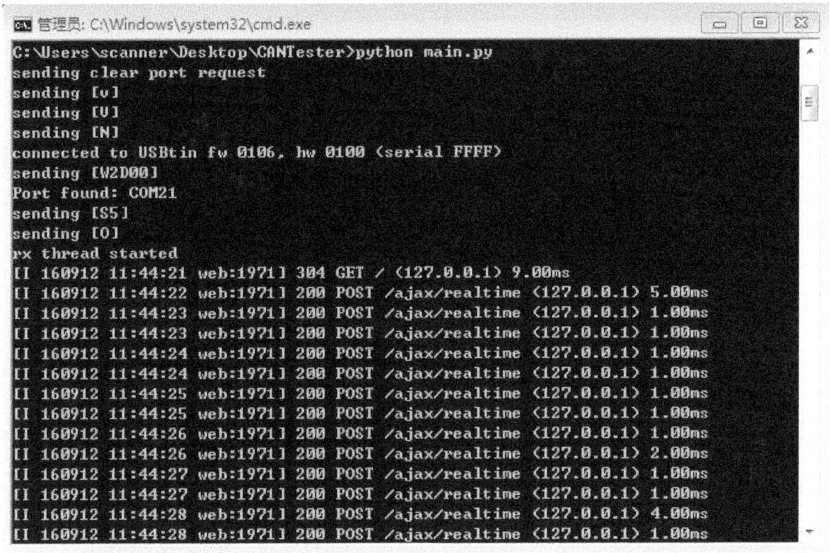

```
管理员: C:\Windows\system32\cmd.exe
C:\Users\scanner\Desktop\CANTester>python main.py
sending clear port request
sending [v]
sending [V]
sending [N]
connected to USBtin fw 0106, hw 0100 (serial FFFF)
sending [W2D00]
Port found: COM21
sending [S5]
sending [O]
rx thread started
[I 160912 11:44:21 web:1971] 304 GET / (127.0.0.1) 9.00ms
[I 160912 11:44:22 web:1971] 200 POST /ajax/realtime (127.0.0.1) 5.00ms
[I 160912 11:44:23 web:1971] 200 POST /ajax/realtime (127.0.0.1) 1.00ms
[I 160912 11:44:23 web:1971] 200 POST /ajax/realtime (127.0.0.1) 1.00ms
[I 160912 11:44:24 web:1971] 200 POST /ajax/realtime (127.0.0.1) 1.00ms
[I 160912 11:44:24 web:1971] 200 POST /ajax/realtime (127.0.0.1) 1.00ms
[I 160912 11:44:25 web:1971] 200 POST /ajax/realtime (127.0.0.1) 1.00ms
[I 160912 11:44:25 web:1971] 200 POST /ajax/realtime (127.0.0.1) 1.00ms
[I 160912 11:44:26 web:1971] 200 POST /ajax/realtime (127.0.0.1) 1.00ms
[I 160912 11:44:26 web:1971] 200 POST /ajax/realtime (127.0.0.1) 2.00ms
[I 160912 11:44:27 web:1971] 200 POST /ajax/realtime (127.0.0.1) 1.00ms
[I 160912 11:44:27 web:1971] 200 POST /ajax/realtime (127.0.0.1) 1.00ms
[I 160912 11:44:28 web:1971] 200 POST /ajax/realtime (127.0.0.1) 4.00ms
[I 160912 11:44:28 web:1971] 200 POST /ajax/realtime (127.0.0.1) 1.00ms
```

Figure 1.15 Start CAN-Pick to connect to the car bus transceiver.

key of fuzzy testing is to formulate test files with effective fuzzy testing rules. For example, the data bits of the CAN data packet are constant, and the CAN ID continuously increases. Researchers can continuously send fuzzing data packets to the CAN bus and observe the response changes of the car. By matching the changes of the car with the data packets sent currently, the CAN message instructions for controlling the car could be found, which is beneficial to reverse the car CAN bus data packets.

Researchers find the CAN ID of the control command, fuzzy test its data bits, and find the bits related to the control command by changing the numbers of the data bits. Then, they use the CAN-Pick software to fuzz the bus. Note that researchers can flexibly formulate the fuzzing rules. Figure 1.15 shows the result of starting CAN-Pick to connect to the vehicle bus transceiver.

After the CAN transceiver is connected, researchers design fuzzy test rules to test the bus. Figure 1.16 shows the interface of fuzzing tool.

The MId field is to set which bits of MId to be fuzzed. The MId fuzzing range field sets the range to be fuzzed. The data field sets which bits of data to be fuzzed. The fuzz mode field selects the mode of fuzzing. The interval field sets the time interval of the fuzzy testing.

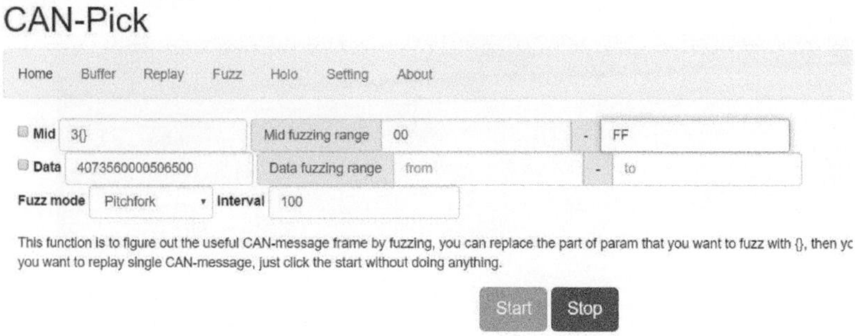

CAN-Pick

Home Buffer Replay Fuzz Holo Setting About

Mid 3{} Mid fuzzing range 00 - FF

Data 4073560000506500 Data fuzzing range from - to

Fuzz mode Pitchfork Interval 100

This function is to figure out the useful CAN-message frame by fuzzing, you can replace the part of param that you want to fuzz with {}, then yc
you want to replay single CAN-message, just click the start without doing anything.

Start Stop

Figure 1.16 The interface of the fuzz testing tool.

1.3 Analysis of OBD-II Interface Attack Technology

1.3.1 Overview of OBD-II

The development of on-board diagnostics (OBD) has gone through two stages: OBD-I and OBD-II. OBD-I, developed by General Motors Company, can only monitor the work of some components and some emission-related circuit failures. Its diagnostic function is relatively limited. Since the data communication protocol for obtaining OBD information and the standard for connecting external equipment interfaces are not unified, OBD-II appeared.

Compared with OBD-I, OBD-II has a greater improvement in the diagnostic function and standardization. The SAE has standardized the communication protocol among the fault indicator light, diagnostic connector, external equipment, and trip computer, as well as the fault code through the corresponding standards. In addition, OBD-II can provide more data that can be read by external devices, including fault codes, real-time data of some important signals and parameters, etc. The advantages of the OBD-II diagnostic system are mainly reflected in the following aspects: first, unite the communication protocol of the car's internal network; second, unite the fault diagnostic interface; third, unite the setting of the fault code; fourth, expand the detection items of the OBD system.

The automobile OBD interface is generally located under steering wheel and near the driver's knee. Its shape is a 16-pin socket as follows, and the pin number of the interface is shown in Figure 1.17.

The automobile fault diagnosis equipment is connected to the vehicle ECU through the 16-pin OBD interface. It sends the service request command to the on-board ECU and obtains the relevant diagnosis data through the

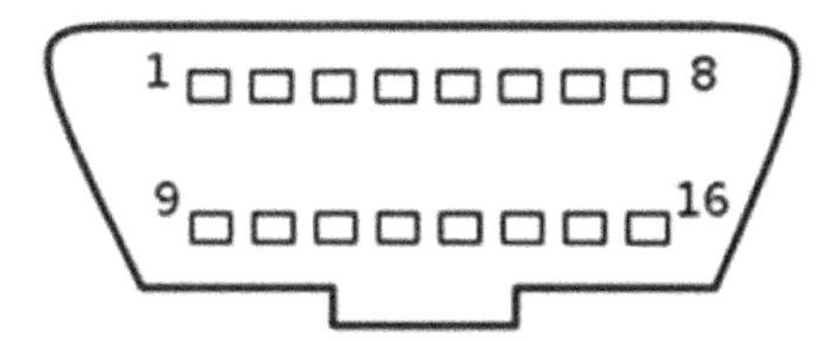

Figure 1.17 The pin number of the OBD-II interface.

ECU acknowledgement message. Though OBD-II has brought great convenience for the vehicle fault diagnosis, OBD-II also has its limitations, mainly reflected in the following two aspects.

First, there is no fault code output. The OBD system cannot diagnose all the faults in the electronic control system. For example, when the power system has open-circuit or short-circuit fault for some reasons, resulting in the engine not being able to start or the vehicle not being able to run normally, the main power of ECU is often in the state of no power; so it is unable to obtain any sensing signals and execute feedback signals, and then the self-diagnosis system cannot work. When the fault indicator and communication interface are damaged, the fault code cannot be obtained. Moreover, the OBD system can only provide "hard" fault codes such as no output signal caused by short-circuit and open-circuit fault or electrical device damage related to electronic control system. When the input and output signal voltage of ECU changes within the specified range, the fault self-diagnosis system will determine whether the electronic control system works normally. Therefore, the self-diagnosis system cannot detect the "soft" faults, such as the deviation of the output characteristics caused by sensitivity reduction and accuracy error of the most actuators and sensors.

Second, the fault code does not necessarily show the real fault location. First, the appearance of fault signal is not only related to the fault of sensor or actuator itself but also to the failure of corresponding wiring circuit failure. When ECU determines a certain circuit fault, it only provides the nature and scope of the fault. It is necessary to further check the wiring, plug, ECU, and related components to determine whether it is the fault of sensor, actuator, or corresponding wiring. Second, the fault code is only a "yes" or "no" definition conclusion recognized by ECU and not necessarily the real fault location. Since the working performance of each component in the system affects each other, resulting in some similar faults, the self-diagnosis system may display false fault codes. Third, when the ECU itself has failures, it will result in the self-diagnosis output signal not being normal and fault code not being able to be used.

Though the application of the OBD system improves the efficiency of fault detection, currently, it still has its own limitations. During the application of the OBD for fault diagnosis, it is necessary to make a comprehensive judgement based on the fault code combined with the information of the fault phenomenon and data flow and further analyse the fault by combining internal self-diagnosis and external diagnosis of the vehicle. With the rapid development of modern electronic technology, microcomputer technology, communication technology, etc., the OBD system will also become more complex, and its on-board detection function will become more powerful and perfect; so it will play a more important role.

1.3.2 Attack Technique Analysis of OBD Box

For automobile users, the OBD equipment is generally divided into two parts, namely OBD box and App. The box is mainly used to collect data on the car, while App can help the car owner detect vehicle faults and record vehicle tracks and violations. In addition to the application based on big data, the later social function is also in the process of further improvement. Figure 1.18 shows the general communication architecture of the OBD box.

Currently, the OBD boxes in the market are generally divided into three categories, that is, Wi-Fi version, SIM card version, and Bluetooth version. The OBD box of Wi-Fi version has a build-in wireless communication module. The OBD box of SIM card version comes with a SIM card. The advantage is that such boxes generally have the security function. When the car doors, windows, or lights are not turned off, or illegal ignition and other situations happen, the mobile phone will automatically alarm. The OBD

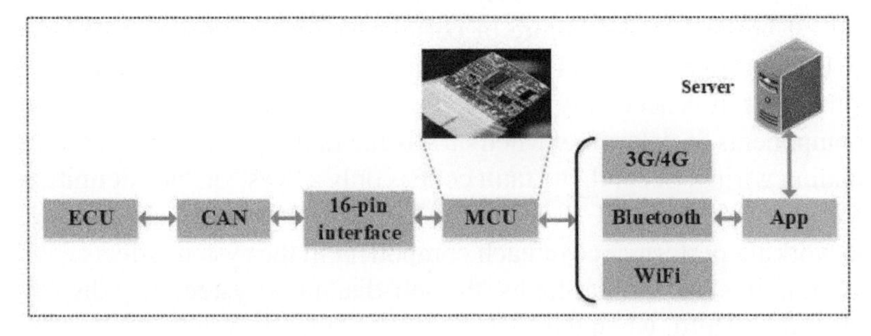

Figure 1.18 The general communication architecture of the OBD box.

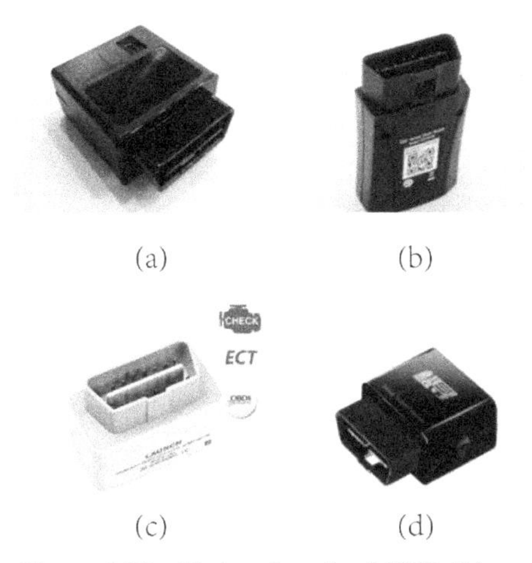

(a) (b)

(c) (d)

Figure 1.19 Various brands of OBD-II box.

box of the Bluetooth version is directly connected via Bluetooth without the security function. Compared with the box of SIM card version with the security function, the price of the box connected with Bluetooth is relatively low. Figure 1.19 shows various OBD boxes.

Considering the life cycle and the communication architecture of the OBD box, the OBD box may have the security threats listed in Table 1.4.

In 2015, Andrew Prudhomme, Karl Koscher, and Tefan Savage from the University of California, San Diego, CA, USA, published a paper "Fast and Vulnerable: A Story of Telematic Failures", which discussed the safety of the OBD box (the researchers referred to it as the car external transmission control unit (TCU)) and how to use the OBD box to control the relevant functions of the car. The following is the analysis of their research results on the OBD box.

1.3.2.1 Attack Surface Analysis

The OBD box analysed by the researchers was purchased from the eBay e-commerce platform, which was manufactured by Mobile Devices Ingenierie and used for insurance purposes. The CPU of this OBD box is an ARM11 chip with a frequency of 500 MHz, whose RAM storage is 64 MB and the flash memory is 256 MB. For the external connection, it mainly includes the following parts: a USB interface, a 2G cellular network data modem (later model 3G), and a CAN transceiver connected to the OBD-II pin.

Table 1.4 Main Security Threats of the OBD Box.

OBD device	Packet injection	Whether the device verifies the received CAN data.
	Identity credential	Whether credentials with which the device verifies visitors written in firmware are immutable or not.
	OBD equipment without security configuration information	Whether the equipment provides accurate security configuration methods and usage precautions.
	OBD equipment hazardous function unknown	Whether the device has functions not written in the manual.
Application	Unreinforced cellphone application	Whether cellphone applications are reinforced.
Equipment and server communication	Firmware update tamper	Whether the device verifies the update downloaded to the local location.
	Firmware update command forgery	Whether the device authenticates the source of the update command.
	Unencrypted communications	Whether the communications between equipment and server are encrypted.
	Weak inspection of HTTPS certificate	Whether the equipment verifies the HTTPS certificate when it communicates with the server equipment via HTTPS.
Equipment and mobile communication	SMS command forgery	Whether the device authenticates the source of the SMS command.
	Wi-Fi insecure encryption	Whether the Wi-Fi communication of the device uses WAP2 to encrypt.
	Wi-Fi weak password	Whether the Wi-Fi initial password of the device is a complex password.
	Bluetooth unused PIN code	Whether the Bluetooth function of the device uses PIN code to authenticate.
	Bluetooth weak password	Whether the Bluetooth initial PIN of the device is a complex PIN code.
Mobile phone and server	Encryption of cellphone communication	Whether the communication between the mobile application and the server is encrypted.
Background server	Management website weak password	Whether the initial password in the background of the device management website is a weak password.
	Management website unencrypted communication	Whether the communication of equipment management website background is encrypted.
	Weak check of background server identity	Whether the background server of the device authenticates the source of the request.
	Background server	Whether the background server of the device is safety reinforced.
Other	No device vulnerability disclosure mechanism	Whether the equipment manufacturer releases information on the device vulnerability.

To evaluate the security performance of this device, the researchers considered two threat models: the local security threat model and the remote security threat model. Local threat model evaluates the security performance by judging whether the attacker can directly attack the OBD box to control the device (for example, gain temporary physical access to the vehicle during transportation). In the remote threat model, it is assumed that the attacker can remotely invade the vehicle's OBD box through wireless attack and then control some functions of the vehicle.

(1) Local attack surface

In the local threat model, it is assumed that the attacker can have physical access to the OBD box. The researchers did not evaluate any communication methods of the vehicle, assuming that an attacker with physical access rights can already directly access the vehicle. The researchers found that the USB interface of the device for debugging can be configured in a network adapter mode (the OBD box becomes a network device at this time). Once the OBD box is connected to the Internet, the web server configured can be used to monitor port 80 and port 23 to detect the information of the OBD box. In addition, the researchers found that the USB interface can be used to make configuration changes and complete software updates. The researchers also analysed multiple test points on the internal circuit board of the device (the package needed to be opened). The attacker can also remove the NAND flash chip and dump or change the contents of the chip through physical access.

(2) Remote attack surface

In the remote threat model, the researchers assumed that the attacker had no physical access to the device or vehicle or even did not know the exact location of the OBD box in the vehicle. The attack target of the remote threat model is the 2G modem that provides a remote connection through a cellular network to remotely control the attacked vehicles. The equipment provides SMS service and data communications based on IP addresses, which are used for various functions of vehicles, and vehicles can be remotely identified, accessed, and controlled.

1.3.2.2 Attack Process Analysis

(1) Local attack analysis

The easiest way to access the device physically is to connect it through a USB port. The OBD box was connected to the network computer by USB cable. By analysing the equipment, the subnet number and IP address of the device could be determined. Then, by utilizing the network interface to detect the service of the device, the researchers found that the OBD box could respond to Telnet, web, SSH, etc., on the standard ports.

(2) Web/Telnet console access

Once connected to the OBD box successfully, both the web server and Telnet server will provide an exclusive interface for querying and setting device parameters and retrieving status information, including some privacy and sensitive information, such as GPS location information. In addition, the researchers detected a set of more advanced instructions, such as instructions to send SMS messages to specified ports. By analysing these instructions, it is possible to directly identify the number in the SMS chip based on the following SMS test. The researchers also determined an interface for retrieving the version and status of all the internal components of the OBD box and the log files of the underlying Linux kernel (both files are helpful for follow-up analysis). Finally, the researchers also found that all configuration variables to run the software module are changeable.

(3) NAND dump

To get more information about the software running on the device, the researchers removed the NAND flash memory chip and used a hardware reader to extract the chip's data. To analyse the extracted data, the researchers created a simulated original NAND flash memory chip exploiting the Linux nandsim kernel module and then configured the module to simulate a copy of the NAND file present on the OBD box. The researchers found an unsorted block image file system (UBIFS), which manages the error detection and correction of the original NAND chip. This file system would record and track the damaged modules.

Researchers used the UBIFS to read the information of the OBD box in partition. One of the partitions contained third-party software running on the OBD box, including a collection of scripts and binary files related to various system operations. The researchers also discovered important information about the transmission and interaction with the OBD box, which contained many public keys, private keys, and certificates.

(4) SSH key

Initially, the researchers could not access the SSH service running on the device. After studying the dump information of NAND flash memory and determining the private key of the root user, this situation changed. The private key of the root user enabled the researchers to authenticate to the OBD box through SSH service and directly obtain the right of the root user of the device. Then, the researchers could read and write any files, download and install other software to create arbitrary functions, and execute arbitrary commands. You might not think it poses a serious threat to cars because an attacker needs to physically access the OBD box device to get the root private key. However, after testing the SSH key on several other devices from the same manufacturer, including the production environment, the researchers found that these devices used the same key, which allowed attackers to control a large number of devices at the same time.

(5) Further analysis

After obtaining the control of the OBD box through the USB interface, researchers focused on how to remotely control the OBD box and further control the car. Two remote communication interfaces attracted the researchers' attention: SMS interface and cellular data interface that provides Internet connection.

(6) Internet-based access

When the researchers check the source code of the OBD box, they found that the web services, Telnet services, and SSH services provided by the OBD box are all bound to the network interface. Since the cellular data modem connected to the Internet is needed for the device to access the Internet, the external attacker can directly use the SSH key previously discovered (IP address known) to login to the box remotely. However, the cellular carriers of most of the OBD boxes that the researchers tested widely used network address translation (NAT) mode; so it was not possible to login to the box directly. The researchers have noticed that the attribute characteristics of cellular devices can be used to access the target devices. If the device uses the services provided by the cellular network provider that allows direct addressing, the direct login to the OBD box will be effective. Later, the researchers found thousands of OBD boxes exposed on the Internet that allowed direct addressing.

(7) SMS-based access

In addition to the cellular data connection, the OBD box also has the ability to receive SMS messages; so if the SMS number of the OBD box is known, outside users can send SMS messages. The SMS management commands that the researchers searched on the network were fully applicable to the tested OBD box. These commands included retrieving or setting the local debugging interface, retrieving the information about the status of the sensors (such as modem information, GPS location, etc.), remote updates, and so on. Among them, the remote update is very dangerous because it may provide a mechanism to obtain a reverse shell (a technique for sending shell commands to remote devices) to achieve arbitrary access to the OBD box.

After checking the log created by SMS initiated update, the researchers confirmed the complete update process of the local files of the OBD box, as shown in Figure 1.20. The update process relies on a special text file UpdateFile.txt from the remote server, which contains the name, path, and hash value of the file to be added or removed from the system. The steps of the update process are as follows.

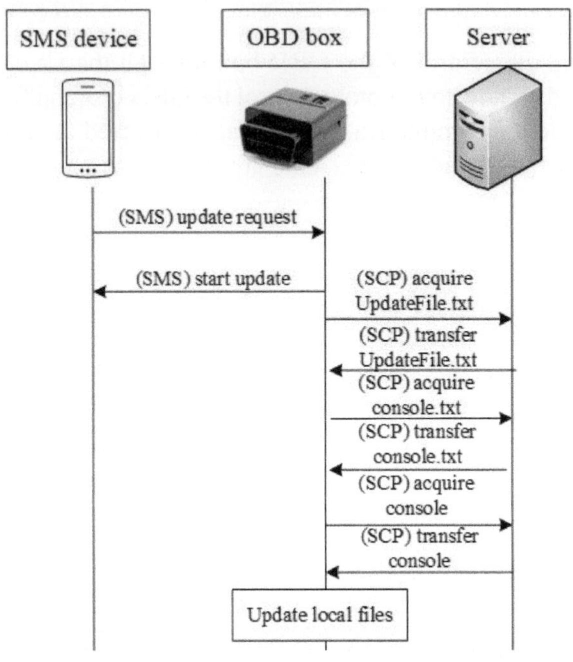

Figure 1.20 Remote update process of OBD box.

• The SMS device sends update information (including user name, host name, port number, path information, etc.) to the OBD box, and the OBD box responds with update and starts.

• Under secure copy protocol (SCP), which is used to define the protocol of transferring files between "the local device and the remote machine" or "the remote machine and the remote machine", the OBD box uses the information sent by SMS device to login to the specified update server and retrieve UpdateFile.txt from the path information.

• Check UpdateFile.txt, retrieve the incorrectly hashed or non-existent files on the local system of the OBD box from the update server under SCP, and send them to the temporary directory of the OBD box.

• If the hash value of the new file is the same as the hash value of UpdateFile.txt, move the new file to the target directory; otherwise, restart the update process.

• If any of the following console commands are included in Update-File.txt, the corresponding action is performed: clear, confirm, reset, etc.

There are many security risks in this update process of the OBD box. First, the update package is not encrypted and signed in any way, which makes it

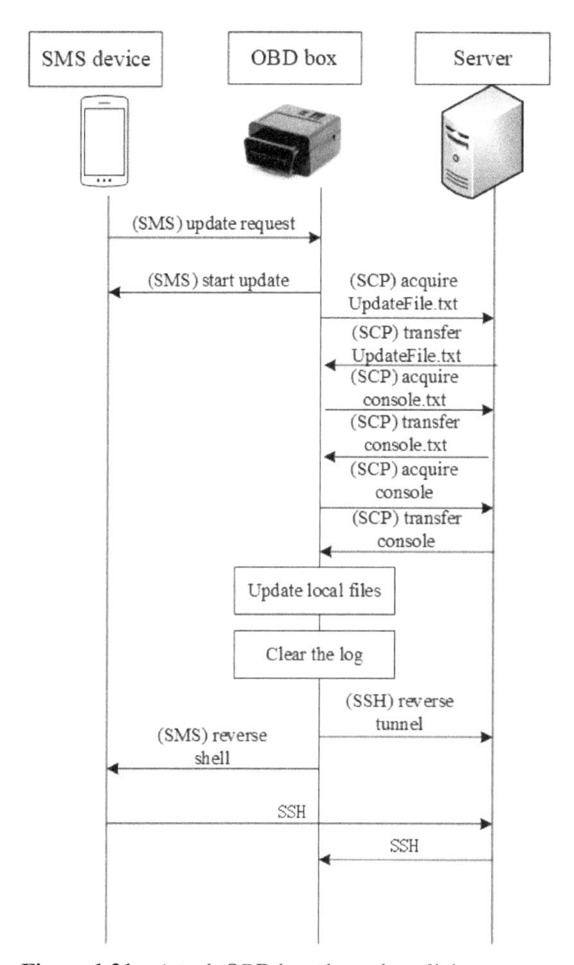

Figure 1.21 Attack OBD box through malicious server.

easy for attackers to replace arbitrary code during the update process. Second, when the server authenticates the device, since all the OBD boxes share the same public key and private key to update the key pair, the OBD box will not authenticate the server.

In order to verify the security risks of the OBD box in the above-mentioned local file update process, the researchers created a malicious update server to provide update for the OBD box and generated a reverse shell to the victim OBD box and a reverse SSH tunnel, as shown in Figure 1.21. The attack process includes the following steps.

• Researchers used the malicious server to initiate remote update through Telnet service, web service, or SMS service.

•The OBD box downloads UpdateFile.txt, which contains *console.bak* (original console binary file), console (shell script containing attack commands), and clear command.

•The OBD box downloads all the files, replaces the system console command with the researcher's console script, and calls "console clear" to clear the log.

• Start the console script and replace it with the original console.bak. Start the reverse shell and SSH tunnel. Send a text message to inform the researchers of the success of the attack.

• Once SMS is received, or a notification from the server that the reverse shell is ready, the researchers can use SSH or tunnel to login to the OBD box to gain root and then control the device or car.

All forms of remote attacks (remote login via SSH, or update via Web, Telnet console, or SMS) require the identification of the OBD box, the globally accessible IP address, or the number associated with the SIM card. The researchers found some ways to find this information.

If the cellular network operator of the OBD box does not provide NAT mode services, the built-in network server provided by the operator can be accessed through the Internet so that the identity of the OBD box can be retrieved. A series of IP addresses of possible OBD boxes can be found by searching special strings on the network. Similarly, vulnerable OBD boxes can be identified according to the information output from Telnet and SSH servers. Since all the OBD boxes of this type use the same SSH server key, the server fingerprints displayed are the same when they are connected. Attackers can search the IP addresses of about 1500 potential devices generated by the fingerprints.

The researchers evaluated the possibility of looking for devices by the number associated with the SIM card. Since the OBD box needs data connection to send sensor information to remote server, the device usually connects with cellular network with prepaid data, and the number required for connection is not random. The researchers found that most of these numbers are assigned sequentially from the 566-area code, which is only used by "personal communication service". If the number of one communication device can be determined, the numbers nearby are very likely to be the numbers of the OBD box.

1.3.2.3 Invade Car CAN Bus

After invading the OBD box, researchers consider how to use the OBD box to control the vehicle. One situation is that the OBD box can only receive CAN data from the OBD-II interface but cannot send CAN messages. The other situation is that the OBD box may allow arbitrary data exchange with the CAN bus, allowing the attacker to directly communicate with every ECU on the vehicle.

The researchers identified two different firmware versions used by the OBD box. The latest version includes the SocketCAN module, which is a Linux kernel module that presents the CAN bus as a network interface. In addition, the OBD box with the new version of the firmware is equipped with the Linux CAN software package. The software package includes several tool software for reading, saving, creating, and replaying CAN messages. Like the packet capture method of traditional network interfaces, any CAN data packets can be sent and received with this tool software.

The earlier version of the firmware designed a custom interface to send commands from the main ARM CPU to the PIC microcontroller used to control the physical CAN controller of the OBD box. After analysing the serial line between the ARM and PIC chip, the researchers analysed enough protocol information in detail and sent their CAN data packets to the PIC chip. However, if the PIC chip detects that the vehicle is not in adaptive cruise control (ACC) mode or engine off mode, it will periodically query the OBD-II interface and will not send CAN data packets.

After reverse engineering of the firmware dumped by the PIC chip, the researchers can identify ACC status and the engine status checks. When both states are "disabled", they need to refresh the firmware of PIC chip again. This series of modifications allows the researchers to send CAN packets to each ECU of the vehicle to control it, regardless of the state of the car.

1.4 Attack Experiment Against Vehicle Bus

In order to verify that there are many vulnerabilities that can be easily exploited in integrity check value (ICV), this part introduces the security test experiment in work [9] and introduces several typical automobile bus attack cases in detail, namely control the display of vehicle dashboard, control the door lock switch, car lights and wiper, as well as CAN bus overload via flood attack. In these attack cases, the working principle of each control unit of the car is introduced in detail, and the corresponding attack mode according to its characteristics is carried out. It should be noted that the introduction of attack

Figure 1.22 Luxgen U5 SUV.

cases is only to make the majority of readers and researchers aware of the importance of the safety in ICVs and seek corresponding solutions according to their loopholes so as to make the future intelligent connected vehicles more secure and reliable.

In the experiment, the test vehicle is Luxgen U5 SUV, as shown in Figure 1.22. Luxgen automobile is not only equipped with a large number of auxiliary intelligent systems such as cruise control system, automatic transmission system, body control system, and automatic parking system but also equipped with remote communication modules such as telematics service provider (TSP) and T-BOX. It is an ideal vehicle for safety test. Figure 1.23 shows the OBD-II diagnostic equipment used in the experiment. Researchers can use CANalyst II diagnostic equipment to collect/send CAN bus data from vehicle OBD-II port.

1.4.1 Control of the Display of Vehicle Dashboard

Automobile dashboard is a device that reflects the working conditions of various systems in the vehicle, and it is also one of the most dependent tools for users in the process of driving. It can display the current state of the vehicle, such as the speed, the engine speed, the temperature of engine, etc. Drivers can adjust the vehicle state according to the driving status information. Therefore, once the dashboard data is tampered with maliciously, it may lead to driver's wrong decision making and even cause catastrophic consequences such as traffic accidents.

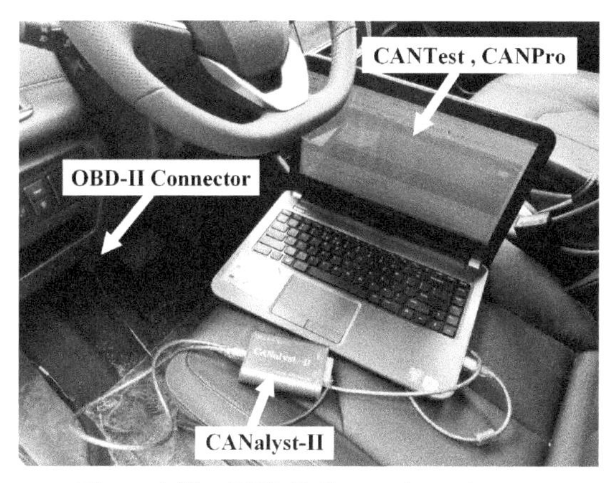

Figure 1.23 OBD-II diagnostic equipment.

In the process of vehicle running, various ECUs used for vehicle speed, engine speed, light, and horn will transmit their status to the receiver of instrument panel through CAN bus. The digital signal processor in the instrument panel analyses the received signal and transmits the decoded data to the display. It is found that the CAN bus in Luxgen vehicle is not equipped with abnormal detection device, and the instrument panel does not carry out relevant safety verification for the identified digital signals. If malicious attackers use CAN diagnostic equipment to capture relevant data and analyse it, and then simulate the information data packet to attack the automobile dashboard, the deception attack to automobile dashboard can be realized.

Through the reverse analysis of CAN bus data, it is found that the data frame with ID of $0x0360$ is related to the speed display in the instrument panel, and the data frame format is: $[X_3X_2\ X_10\ 00\ 00\ 00\ 2Y\ 00\ MN]$. Among them, $X_3X_2\ X_1$ represents the current speed, with the value ranging from $0x000 \sim 0x5DA$ and the speed ranging from 0–199 km/h. Y is related to pressure sensor, and MN is a counter that reflects the sequence of vehicle speed. According to the data format and rules, the actual speed of the vehicle can be calculated by using the following formula:

$$Kilometersperhour(kph) = 34.5 * X_3 + 2 * X_2 + X_1/8. \tag{1.1}$$

After cracking the vehicle speed code in CAN bus, a script can be written to generate a group of CAN control commands, which are injected through

the OBD-II port to realize the deception of the instrument panel. Some attack data injected are as follows:
SEND 0x0360 DATA Frame 8 17 10 00 00 00 20 00 AE
SEND 0x0360 DATA Frame 8 17 00 00 00 00 20 00 B0
SEND 0x0360 DATA Frame 8 17 10 00 00 00 20 00 B1
SEND 0x0360 DATA Frame 8 17 00 00 00 00 20 00 B2
SEND 0x0360 DATA Frame 8 17 10 00 00 00 20 00 B3
SEND 0x0360 DATA Frame 8 17 00 00 00 00 20 00 B4.

Here, "17 0" means that the current speed displayed by the instrument cluster is $34.5\times1+2\times7+0/8 = 48$ km/h. During the sniffing analysis, it was found that the manufacturer of Luxgen added counters at the end of the data to enhance the security of CAN data. However, in the attack experiment, the experimental attack is still effective when the speed display of the instrument panel is directly attacked by abandoning the analog counter.

In addition, the researchers also used the same attack methods to control the display of engine speed in the instrument panel. After analysis, it is found that the data frame with ID of $0x$ 0316 in CAN bus is related to the engine speed. The format of engine speed packet is as follows:

$[01\ Y_1Y_2\ X_1X_2\ 6C\ Y_1Y_2\ 40\ 16\ 62]$.

The value range of X_1X_2 is $00 - F0$, which means that the range of engine speed is $0 - 7800$ rpm. Y_1Y_2 is the size of the pressure sensor on the car throttle, and its value range is $3E - 7F$.

Through in-depth study of engine speed, the value can be obtained by the following formula:

$$\text{RPM} = (5.2 * X_1 + X_2/3) * 100. \tag{1.2}$$

According to the above research conclusions, scripts are written to automatically generate attack data about engine speed in the instrument panel. Some of the generated data are as follows:
RECV 0x0316 DATA Frame 8 01 5D 3B E3 5D 38 47 3D
RECV 0x0316 DATA Frame 8 01 5D 3C 48 5D 38 47 3D
RECV 0x0316 DATA Frame 8 01 5E 3C AF 5E 38 47 3D
RECV 0x0316 DATA Frame 8 01 5E 3C AF 5E 38 47 3D
RECV 0x0316 DATA Frame 8 01 5E 3D 21 5E 38 47 3D
RECV 0x0316 DATA Frame 8 01 5D 3D 87 5D 38 47 3D.

From the data frame part of this attack data, $3C$ indicates that the engine speed displayed is $(5.2 * 3 + C/3) * 100 = 1960$ rpm. Note that in the same data frame, the data of the second byte and the fifth byte are the same, which is the security verification mechanism of CAN bus.

Figure 1.24 The experimental results of malicious tampering in automobile dashboard.

In the automobile dashboard, lights, engine temperature, safety belt warning, door switch status, and other display contents can be controlled. In the process of driving, if the dashboard display is maliciously controlled, the wrong display will mislead the driver's judgement, thus leading to the occurrence of dangerous situations.

Figure 1.24 shows the experimental results of maliciously tampering in the dashboard. When the car is moving, a malicious attacker sends attack instructions to the vehicle via its OBD-II port. It can be seen from the figure that the vehicle speed displayed on the instrument panel is 161 km/h, the engine speed is 7400 rpm, the current engine temperature is 0, and the car lights and seat belts are in warning status. Under normal conditions, these values are unlikely to occur.

1.4.2 Vehicle Status Tampering via Body Control Module

Nowadays, in most vehicles, the body control module (BCM) plays the role of vehicle gateway, which is responsible for controlling the functions related to the body (such as headlight, wiper, window, door lock, and body stability). BCM is also responsible for coordinating the communication between different network segments in the car, which solves the problem of complicated wiring between ECUs. It reduces the cost of automobile manufacturing and the failure rate of vehicles. Therefore, the BCM has gradually become an integral part of vehicle design. However, while providing convenient service for vehicles, BCM also has a great potential safety hazard; for example, using the loopholes of BCM to control the door lock switch, the car light, wiper, etc.

1.4.2.1 Control the Door Lock Switch

Intelligent remote control key system has been widely used in today's intelligent connected cars. The user can control the door switch, trunk switch, and horn honking within a certain distance through the key. When the driver uses the remote key to control the door, the key will send control instructions and security verification code to the key control unit (KCU). KCU will verify its ID code after receiving the input signal. If the ID number is approved, KCU will feed back an authentication signal to the key and send the door control signal to the BCM via CAN bus. After receiving the control signal from KCU the BCM controls the door lock status according to the command. The interactive authentication process between smart remote key and vehicle is shown in Figure 1.25. The perfect ID code authentication mechanism

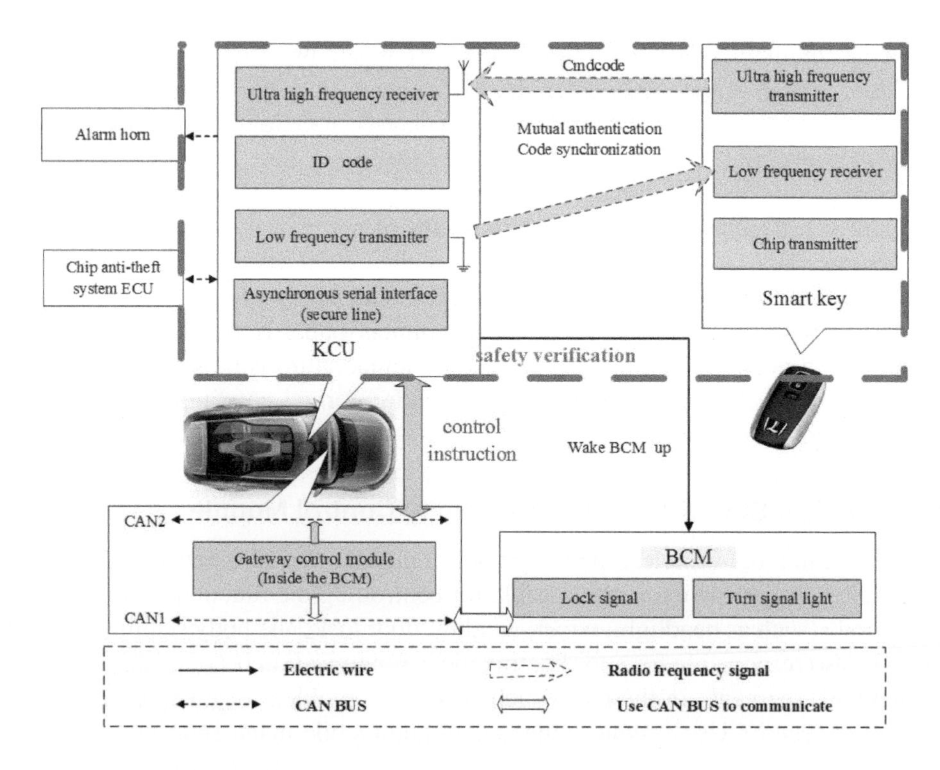

Figure 1.25 The working principle of automobile remote key.

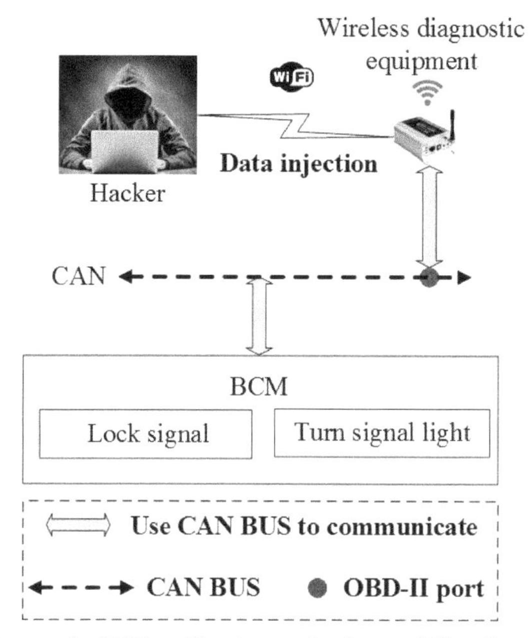

Figure 1.26 Bypass the KCU verification mechanism and illegally control the door lock switch.

between key and car makes it difficult for malicious attackers to forge radio attack signals to invade the car.

The researcher analysed the working principle of the intelligent remote control key system in detail, taking advantage of the vulnerability of no security check between KCU and BCM. They bypassed the security verification mechanism between the key and KCU, and successfully captured the door lock control instruction between KCU and BCM from the CAN bus. Then, they directly control the door switch from the CAN bus by using the door lock control instruction. The working principle of attacking car door locks is shown in Figure 1.26.

1.4.2.2 Car Lights and Wiper Attacks

It is found that the BCM does not provide corresponding security protection for data when communicating with other ECUs in CAN bus. When the driver controls the car lights and wipers, the light/wiper combination switch module will send instructions to the BCM via CAN bus. After receiving the command signal, BCM will directly send the control signal to the actuator to control the

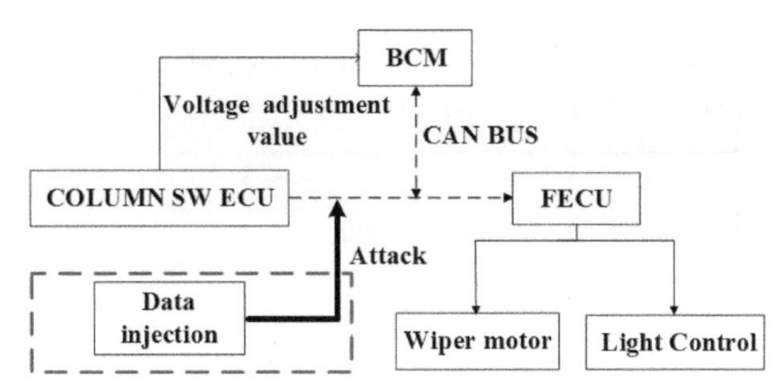

Figure 1.27 Bypass the BCM verification and control the car lights and wipers from the CAN bus.

work of the car light/wiper. Figure 1.27 is the working principle diagram of the light/wiper control by the hacker bypassing the automobile BCM control module. Hackers steal packets sent by BCM in the same way on the CAN bus and study the instructions in them. The camouflage data package is sent back to CAN bus, which is transmitted to the actuator to control the work of lights and wipers.

The ID associated with vehicle lights and wipers is $0x$ 0410, and its data format is as follows:

$[X_1X_2\ X_3X_4\ Y_1Y_2$ 00 00 00 00 00$]$.

Among them, the $X_1X_2\ X_3X_4$ is used to control the speed of vehicle wiper. Y_1Y_2 is used to control car lights. In order to make the research more interesting, the researchers used the music in vehicle to automatically generate the attack command in the experiment and injected it into the CAN bus to realize the function of controlling the car lights and wipers.

1.4.3 CAN Bus Overload via Flood Attack

CAN bus uses priority strategy to transmit data. In CAN bus, when multiple ECUs send messages at the same time, the smaller the CAN ID number is, the higher the priority of data transmission in CAN bus. According to this feature, attackers can send a large number of packets with ID $0x$0000 to CAN bus, which overloads CAN bus data and causes it to refuse service. When other ECU cannot receive and send messages, the car will be paralysed.

This attack method takes advantage of the inherent characteristics of CAN bus to carry out flood attack. It should be noted that this attack method is very dangerous, especially when the car is moving. What is more, this attack method can cause damage to the vehicle. Therefore, readers are not advised to imitate.

Figure 1.28 Experimental vehicles: Luxgen U5 SUV and Buick Regal.

1.5 Experiment of Driving Behaviour Analysis

Technology is a double-edged sword, which can not only be used to achieve malicious attacks in vehicles but also can be used to do something meaningful. In this section, a new idea will be introduced, which is to extract driving habits from driving data in CAN bus and realize illegal driver detection, legal driver identification, and driving behaviour evaluation [10]. Nowadays, there are still many valuable works in ICVs to be explored, and with the progress of science and technology, the cars will be more intelligent.

As an old saying goes, there is a thousand Hamlets in a thousand readers' eyes. Driving behaviour is the same. Different people drive the car in a different way. One has his or her own preferences about how they hit the gas and brake pedals and turn the steering wheel, and the distance they follow or run parallel with another vehicle. These habits transmitted into CAN bus to generate data and actuator controls the car according to the data. So, researchers harness CAN data to refine characteristics to analyse driving behaviour. In the experiment, Luxgen U5 SUV and Buick Regal are experimental vehicles, as shown in Figure 1.28.

1.5.1 A Novel Scheme About Driving Behaviour Extraction

There are three main parts in this subsection. First, researchers introduce how they collect and process vehicle data. In the process, a sliding window was designed to make sure that deep learning (DL) can deal with these data. The regulation of how to generate a label was explained in the second part. In

Table 1.5 Data Form in CAN Bus.

Direction	Time	ID	Data
RECEIVE	10:49:29	0x03f0	**A0** 00 **10** 00 00 00 00 00
RECEIVE	10:49:29	0x03f0	**82** 00 00 00 00 00 10 92
RECEIVE	10:49:30	0x0410	03 **FF** 00 00 00 00 **3C** 01
RECEIVE	10:49:31	0x03D2	44 54 01 **B3** 00 00 00 **0F**
RECEIVE	10:49:31	0x04F3	**FF** 40 62 **08** 00 01 20 00
RECEIVE	10:49:31	0x059A	52 **47** B3 19 0E **F2** 00 00
RECEIVE	10:49:21	0x03DA	01 4F 18 **04** 4F 42 16 **62**
RECEIVE	10:49:32	0x0431	40 **10** 00 00 02 04 **A2** 00

the third part, the details of DL models' inner construction and their specific parameters are introduced.

1.5.1.1 Extraction and Processing of Driving Data

1) Data Extraction and Processing. On-board diagnostic system II (OBD-II) is an interface embedding with automotive diagnostic equipment that has data record and diagnosis function. CAN data is transmitted in the in-vehicle network in the form of broadcast and the researchers collected data from OBD-II. However, there is not only the driving-behaviour-related data but also some irrelevant data. Researchers need to extract useful data from all these multifarious data to improve the accuracy of experiment. Table 1.5 shows the form of data in OBD-II. During movement of car, red bold data keep changing all the time while the other part data stay the same. So, they extract the red bold data as driving behaviour characteristic data and the number of characteristics is m.

2) Characteristic Data Preprocessing. After driving behaviour characteristic data (m characteristics) are collected, they preprocess the data to a format that DL can cope with. There are two steps in this part. First, as Table 1.5 shows, the data are hexadecimal which need to be converted to decimal. Second, every ECU in vehicle sends signal to OBD-II in the name of several concrete IDs. The frequencies of IDs in OBD-II are different; the smaller the ID number is, the higher priority it possesses, and the more frequent the transmission data is. So, they need to unify the frequency of different characteristic data. The average value of each characteristics data is calculated in one second as a recognition unit, leading to a m columns data matrix.

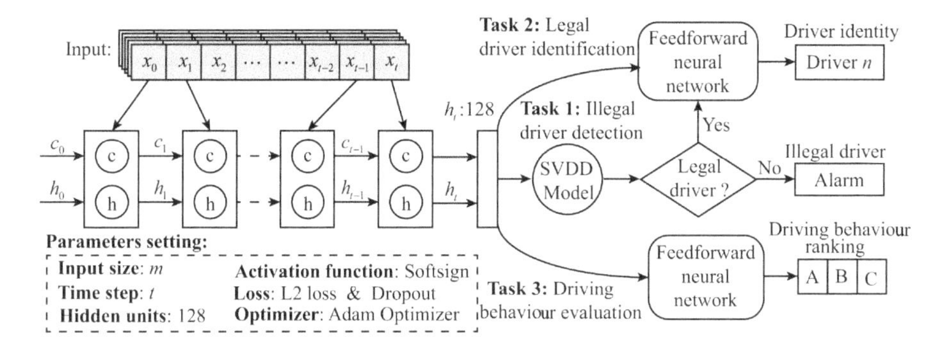

Figure 1.29 MTL network structure. Illegal driver detection, legal driver identification, and driving behaviour evaluation share the same data source and similar feature space.

3) Sliding Windows. Data normalization is a necessary step before conducting an experiment. The whole framework of MTL is shown in Figure 1.29. Input of MTL is a matrix with shape of $length \times width$; so the training/testing data should be sampled in fixed-width sliding windows. In MTL, the length of sliding window corresponds to the time series t, and they set $t = 120$ empirically. The width of sliding window is m. So, the final format of input in MTL is a matrix with $120 \times m$ dimension. Slide window slides once every second for better capability.

1.5.1.2 Grade Regulation and Label Design

The scheme needs an amount of people to be drivers and passengers. Passengers need to grade drivers about their driving behaviour from 1 to 10 score according to certain regulations, and the final score of a driver is the average value represented by s. The authors divide the 10 scores into three different categories: A, B, and C. $0 \leqslant s < 6$ is C level (fail), representing poor driving behaviour giving passengers a bad experience, and $8 \leqslant s \leqslant 10$ is A level (merit), which is the opposite side of C level. $6 \leqslant s < 8$ score is B level (pass), which means the behaviour is not bad but also not so good. The novelty of their scheme is that they make regulations about driving behaviour judgement so that passengers can score the driver's behaviour based on it. There are five rules on how to evaluate driving behaviour and they set 2 scores for every regulation. The rules are described as follows.

- Get 1 score if drivers launch the car slowly and stably, and 0 if they start the car with a big acceleration getting customer a jerk. At the end of driving, if drivers suddenly hit the deceleration pedal to stop the car making the customer uncomfortable, the score is 0. If the stop process is smooth and steady, the score is 1. Only the driver who starts and stops the vehicle well can get 2 scores.

- If the driver always steps on the accelerator/decelerator pedal suddenly, making passengers feel unsafe or anxious, they will get 0 point. 1 score means drivers accelerate/decelerate not so often or do it suddenly, and 2 scores represent that the driver can speed up or slow down the vehicle smoothly, and the driving speed is comfortable for passengers.
- Drivers get 1 or 2 scores if the passenger just feels a little or no jolt when the car goes through the speed bumps or a period of rugged road, and 0 if the jolt is too much. This might be a little subjective, but the average score of all passengers is adopted; so the influence can be ignored in study.
- If drivers take a bend much too fast and turn the steering wheel too hard, the lateral acceleration can be big, which will make passengers wag from side to side. Unbalance can be felt directly by passengers and can change their experience. If passengers cannot control their body wag to another side while drivers take a bend, then the drivers get 0 score. If passengers get no influence of inertia and can sit comfortably, the drivers get 2 scores. Otherwise, drivers get 1 point.
- The cornering lamp is important while driving as it gives signal to other vehicles when switching lane or taking a bend. So, it is adopted as a rule to evaluate the driving behaviour. If drivers light the correct cornering lamp every time they want to change a lane and turn around, they get 1 score, otherwise 0. If drivers fasten seat belt all the way, they get 1 otherwise 0. Only do both two things, they can get 2.

Before passengers get on the vehicle, they are all notified about the regulations. The authors give them handouts with printed regulation. After the trip, the passengers give a score after every rule. The authors sum them up and calculate the average score of every driver and then classify all drivers into three levels A, B and C as the label of behaviour evaluation, which is important in computing accuracy in experiment.

1.5.1.3 Establishment of MTL Network

Multi-task learning (MTL) framework is illustrated in Figure 1.29. The characteristics of driving behaviour in our study are time-dependent. Long short term memory (LSTM) can remember cell states long time ago; so the authors choose it as framework foundation of their supplement experiment.

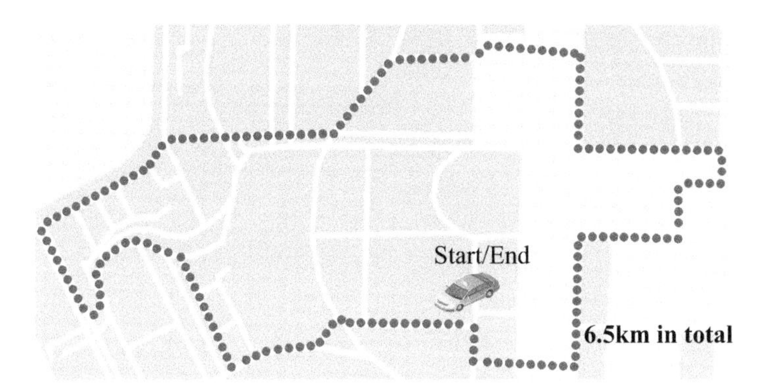

Figure 1.30 Experimental road map. Each driver makes six circles around a fixed route.

According to the number of characteristics $n\_input = m$, length of input corresponds to time series t. Adam optimizer is harnessed to optimize parameters automatically during training. To prevent data over-fit in experimental result, they use L2 normalization and Dropout layer mechanism. L2 normalization is the square root of quadratic sum of all elements in weight matrix. Dropout mechanism is to drop a layer with a probability to prevent over-fit. Softsign activation function is chosen to better utilize the nonlinear factors in MTL.

It can be seen from Figure 1.29 that after extracting the common hidden features of the three tasks by LSTM, support vector domain description (SVDD) algorithm is used to illegal driver detection task. If the driver is identified as a legal driver, MTL model will use feedforward neural network to detect his/her identity and judge the driving behaviour level. If the driver is judged to be illegal, the alarm will sound and the owner shall be informed.

1.5.2 Experimental and Numerical Results

1.5.2.1 Volunteer Recruitment and Experiment Route

The authors recruited 45 drivers and 12 passengers in the experiment. The drivers use two cars along the designated route for 6 laps. The route map is shown in Figure 1.30. The passengers took the car and scored the driver according to the scoring rules, and the average score of the driver will be used as a label for driving behaviour evaluation.

1.5.2.2 Illegal Driver Detection Results

Table 1.6 shows the confusion matrix of illegal driver detection based on Luxgen and Buick. It can be seen that in Luxgen U5 SUV, the accuracy rate of

Table 1.6 Confusion Matrix of Illegal Driver Detection in Luxgen U5 SUV/Buick Regal (%).

Actual labels	Prediction results	
	Authorized driver	**Unauthorized driver**
Authorized driver	**90.5%** **90.9%**	9.5% 9.1%
Unauthorized driver	4.5% 5.3%	**95.5%** **94.7%**

Table 1.7 Performance Index and Results of Illegal Driver Detection in Luxgen U5 SUV/Buick Regal.

Evaluation index	Computational formula	Results	
		Luxgen	Buick
A_i	$A_i = \frac{T_P+T_N}{T_P+T_N+F_P+F_N}$	0.93	0.928
P_i	$P_i = \frac{T_P}{T_P+F_P}$	0.952	0.945
R_i	$R_i = \frac{T_P}{T_P+F_N}$	0.905	0.909
$F_1 - Measure_i$	$F_1 - Measure_i = \frac{2 \times P_i \times R_i}{P_i+R_i}$	0.928	0.927
M_i	$M_i = \frac{F_N}{T_P+F_N}$	0.095	0.091
F_i	$F_i = \frac{F_P}{T_N+F_P}$	0.045	0.053
AUC$_i$	Refer to [2, 4]	0.952	0.947

authorized drivers being correctly identified is 90.5% and that of unauthorized drivers is 95.5%. In Buick Regal, the correct identification accuracies of authorized and unauthorized drivers were 90.9% and 94.7%, respectively. The experimental results show that our MTL model has high recognition accuracy for illegal driver detection.

In addition, in order to verify the excellent performance of their model in illegal driver detection task, they introduce several evaluation indexes. The detailed evaluation indexes, calculation formula, and experimental results are shown in Table 1.7.

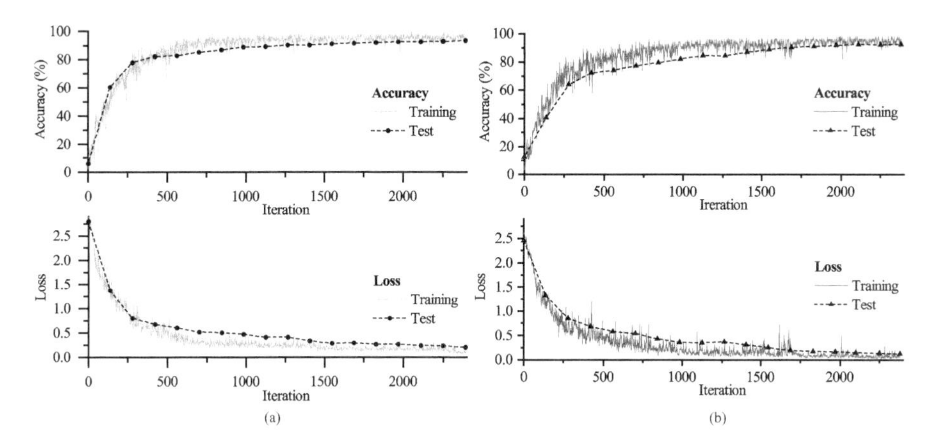

Figure 1.31 Accuracy and loss curves of legal driver identification in Luxgen U5 SUV and Bucik Regal. (a) Legal driver identification results in Luxgen U5 SUV. (b) Legal driver identification results in Buick Regal.

1.5.2.3 Legal Driver Identification Results

Figure 1.31 describes the accuracy curves and loss curves of legal driver identification in Luxgen and Buick cars. It can be seen from the figure that the training accuracy rate of the Luxgen vehicle is 94.9%, and the loss rate is 0.156. The testing accuracy rate is 93.5%, and the loss rate is 0.204. In Buick Regal, the accuracy of training curve and that of test curve are 95.6% and 93.7%, respectively, and the loss rate is 0.114 and 0.185, respectively. These experimental results prove that the driver's identity can be identified with high accuracy by using the driving data, and their model has excellent performance and is without over-fitting.

1.5.2.4 Driving Behaviour Evaluation Results

The authors can also rely on driving data to judge whether the driver has good or poor driving behaviour. Driving behaviour evaluation has high commercial value. For example, bus/taxi companies can use it to recruit drivers with better driving behaviour, and private owners can use it to improve their driving behaviour, which can effectively reduce traffic accidents. For this reason, they rely on Luxgen and Buick to verify the performance of our MTL model. The experimental results are shown in Table 1.8 and Figure 1.32.

The confusion matrix of driving behaviour evaluation in two experimental vehicles is shown in Table 1.8. In Luxgen vehicle, the accuracy rates of driving behaviour levels A, B, and C were 92.76%, 94.89%, and 98.17%,

Table 1.8 Confusion Matrix of Driving Behaviour Evaluation in **Luxgen U5 SUV/Buick Regal** (%).

Passenger rating label	Prediction results		
	A	B	C
A	**92.67%**	6.5%	0.83%
	94.4%	4.67%	0.93%
B	3.61%	**94.89%**	1.5%
	6.47%	**93.07%**	0.46%
C	0	1.83%	**98.17%**
	1.12%	1.1%	**97.78%**

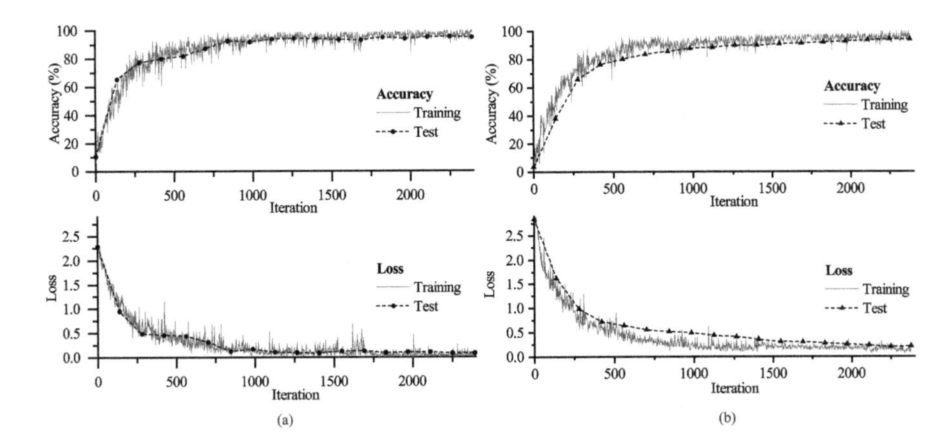

Figure 1.32 Accuracy and loss curves of driving behaviour evaluation in Luxgen U5 SUV and Buick Regal. (a) Driving behaviour evaluation results in Luxgen U5 SUV. (b) Driving behaviour evaluation results in Buick Regal.

respectively. In Buick Regal, the accuracy rates were 94.47%, 93.07%, and 97.78%, respectively. Experiments show that drivers with poor driving behaviours have the highest recognition accuracy, that is, their bad driving behaviours are more easily distinguished by MTL network.

The accuracy curve and loss curve of driving behaviour evaluation accuracy in the two experimental vehicles is shown in Figure 1.32. It can be seen from the figure that the training accuracy rate of the Luxgen vehicle is 98.5%, and the loss rate is 0.029. The testing accuracy rate is 95.1%, and the loss rate is 0.076. In Buick Regal, the accuracy of training curve and test curve are 97.8% and 94.3%, respectively, and the loss rate is 0.139 and 0.168, respectively. The experimental results show that the driving behaviour evaluation scheme can identify the driver's driving behaviour level with high accuracy and without over-fitting. They hope that the relevant research can be gradually improved and applied to the future intelligent vehicles, which can effectively reduce the occurrence of traffic accidents.

2

Intra-Vehicle Communication Security

2.1 Basic Introduction to Intra-Vehicle Communication

2.1.1 Overview of Intra-Vehicle Communication

Intra-vehicle networking and communications can realize the transmission of status information and control signals among sensors, actuators, and electronic units of vehicles, combine the wired and wireless technologies to form an extensible connection backbone structure, and actualize advanced functions such as state perception, motion control, and fault diagnosis in a centralized system.

Automobile short-range communication technology is evolved from contactless authentication and interconnection technology. With the wide application of multiple short-range communication technologies, vehicles need to deploy multiple short-range communication modules to implement various functions such as Wi-Fi, Bluetooth, radio, tire pressure monitor system (TPMS), and keyless system. The security threats faced by short-range communication mainly arise from communication protocol, communication process, communication module, etc. These will specifically involve reverse protocol cracking, sniffing and eavesdropping, data destruction, man-in-the-middle (MITM) attacks, denial of service (DoS) attacks, etc.

Automobile remote communication includes communication between the vehicle-mounted terminal and the cellular network, communication with the mobile terminal, and that with other vehicles and roadside units. Since it needs to meet the requirements of different application scenarios for communication and data exchange, there are also various security threats to communication, such as signal sniffing, MITM attack, and replay attack, which cannot guarantee data confidentiality, integrity, and communication quality. As for the off-vehicle communication threats, they mainly involve the Internet of vehicles TSP platform, remote car control app, over the air (OTA) remote upgrade, global positioning system (GPS) communication, etc.

2.1.2 Module Threat Analysis

2.1.2.1 Threat Analysis of Keyless Entry

The keyless entry system is not based on a traditional key or does not require keys at all. It can be divided into two types: remote keyless entry (RKE) system and passive keyless entry (PKE) system. In the RKE system, there is no need for people to conduct a series of actions of touching the key with their hands, inserting the key into the lock hole, and turning the key. They can unlock or lock the vehicle by simply pressing the corresponding button.

The RKE system is mainly composed of a receiver controller installed on the vehicle, and a transmitter carried by the user, that is, a wireless remote control door key. It allows users to use the transmitter on the keychain to lock and unlock the car door. The data can be transmitted to the vehicle, and the user can trigger the system to work by pressing the button switch on the keychain.

The appearance of the PKE system further omits the operation of pressing the key button. On the premise of carrying with a legal PKE key, the system automatically recognizes the key when the person approaches the vehicle to a certain extent. Then, according to the user's operation, such as touching the door handle, corresponding unlocking or locking operation can be performed, which further improves the convenience.

The threats faced by the keyless entry system are as follows. First, an attacker can steal the user's wireless key signal via signal relay or signal replay and send it to the smart vehicle to deceive it to unlock. Such replay attacks are restricted by number of times. Second, an attacker can look for the algorithm vulnerability of the car key authentication communication to attack. This kind of attack against the authentication algorithm is permanently effective. For example, security flaws have been revealed in the rolling code chip of HCS series and the Keeloq algorithm. The system will determine the signals that meet certain conditions as effective and unlock the vehicle.

2.1.2.2 Threat Analysis of TPMS

TPMS is a common automotive electronic control system used to monitor the air pressure in the pneumatic tires of vehicles. TPMS needs the tires to be equipped with sensors capable of signal transmission. In addition to sensors, signal receivers are required to be installed on the vehicle. By receiving data packets of tire pressure monitoring, the tire sensor ID can be obtained. Then, based on the ID of the tire pressure sensor, the corresponding data is filtered to obtain the tire pressure value of each tire. Finally, the real-time tire pressure information is reported to the vehicle driver through the dashboard or simple low-pressure warning light.

TPMS mainly has the following threats. First, an attacker can listen to and replay the tire pressure signal. Once the tire pressure data packets are received, the ID value of the tire sensor can be decrypted, and the tire pressure value can be tampered, which, in turn, triggers the TPMS to sound the alarm. Second, the uniqueness of the sensor ID can be exploited to track the vehicle. The TPMS signal receiver is placed in the target tracking area, and the TPMS sensor is used to track the vehicle.

2.1.2.3 Threat Analysis of Wi-Fi

Wi-Fi is a wireless local area network (WLAN) technology based on the IEEE 802.11 standard. On-board Wi-Fi requires intra-vehicle Wi-Fi equipment, which generally refers to the technology of transferring 3G/4G/5G to Wi-Fi that is loaded on vehicles and provides wireless routing of Wi-Fi hotspots. The Wi-Fi module is generally integrated in the in-vehicle infotainment (IVI) of vehicles, as shown in Figure 2.1. Consumers can not only use various smart mobile devices but also update the software in the vehicle's Wi-Fi environment.

The current on-board Wi-Fi mainly faces the following threats. First, an attacker can connect the on-board Wi-Fi to intra-vehicle network and attack the intra-vehicle network module, such as IVI. Second, the default password of on-board Wi-Fi is relatively simple, and an attacker can sniff passwords and attack users on the same local area network after accessing the intra-vehicle network. Third, on-board Wi-Fi traffic can be maliciously exhausted so that it cannot work properly.

2.1.2.4 Threat Analysis of Bluetooth

Bluetooth is a wireless communication technology. The Bluetooth communication technology in the on-board Bluetooth system continues from the traditional one. The main application of on-board Bluetooth is to use Bluetooth technology to connect with a mobile phone while the vehicle is in motion to achieve the purpose of liberating both hands and reducing hidden dangers of traffic accidents. Bluetooth technology is also used for car keys or combined with mobile apps to control vehicle doors and windows within a certain range.

The security threats that exist in on-board Bluetooth include the following two parts. One is that if an old Bluetooth version is used on board,

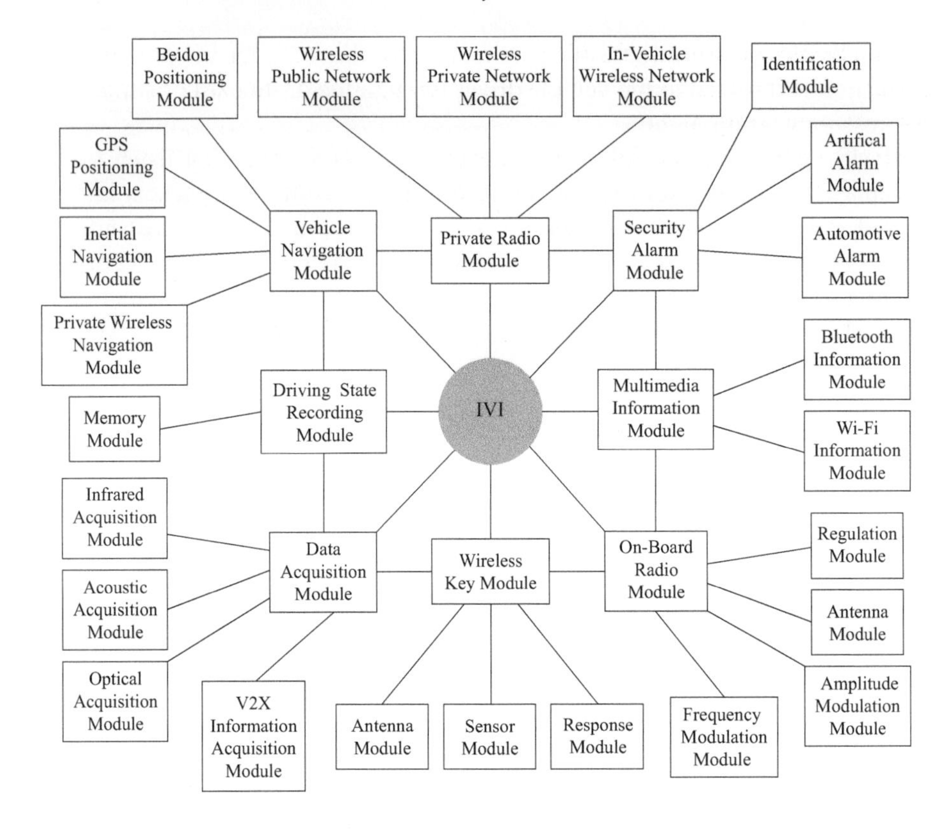

Figure 2.1 In-vehicle infotainment.

there are a lot of security flaws, such as the personal identification number (PIN) code is vulnerable to eavesdropping attacks. If enough time is given, the eavesdroppers can collect all the paired frames and crack the long-term key (LTK). Consequently, they can crack the ciphertext encrypted by advanced encryption standard-counter mode with cipher block chaining message authentication code (AES-CCM) in the Bluetooth link layer, obtain the authorization code, universally unique identifier (UUID) and other data in the data packets, and attack the vehicle user or the vehicle's Bluetooth-enabled devices. The other is the security risks caused by improper encoding when the Bluetooth-enabled devices implement the Bluetooth function. For instance, CVE-2017-0781, a serious vulnerability of the Android Bluetooth stack, allows an attacker to remotely obtain the command execution permission of the Android phone.

2.1.2.5 Threat Analysis of FM

The main performance indicators of on-board radios are frequency range, sensitivity, selectivity, overall frequency characteristics, overall harmonic distortion, output power, etc. Frequency range refers to the frequency of radio broadcast signals that the radio can receive. To listen to the radio, the radio waves must be first captured. The receiving antenna of the radio is used to receive radio signals.

The wireless communication based on the radio mainly faces the following threats. The first one is wireless data packet monitoring. Monitoring equipment with the same operating frequency as the target system is used to collect and reversely analyse wireless data packets. After corresponding decryption of wireless data packets, the operation principle of the entire radio system can be deeply understood, and, thus, the key commands can be found to implement remote control of vehicles. The second is the wireless signal spoofing attack, which combines the wireless monitoring and the decryption method, parses the wireless communication protocol, and constructs a legally authenticated wireless packets to spoof the target. Third, the wireless signal hijacking attack uses the method of wireless protocol layer or communication layer network blocking to hijack the wireless signal.

2.1.2.6 Threat Analysis of GPS

GPS is a satellite navigation system developed by the US Department of Defense. With all-round, all-weather, all-time and high-precision, it can provide low-cost, high-precision navigation information such as position, speed, and precise timing. GPS vehicle application systems are generally divided into two categories. One is that the vehicle tracking system used for anti-theft of vehicles. Since a major feature of the GPS receiving system is to only accept but not transmit signal, the GPS tracking system provides the vehicles with anti-theft services with the support of communication networks and auxiliary systems. The other is for autonomous navigation of vehicles. The vehicle navigators timely grasp their own position and destination by receiving satellite signals and cooperating with electronic map data.

The security threat of GPS is that it is vulnerable to be interfered and deceived. The GPS satellite broadcasts signals continuously to tell the vehicle where the GPS is. The satellite signal is a unidirectional broadcast signal, and after the long-distance transmission from space to the ground, the signal has become very weak. A GPS simulator can be disguised as a satellite next to the receiver, the signal of which can easily cover up the real GPS signal and, consequently, deceive the GPS receiver module of vehicle.

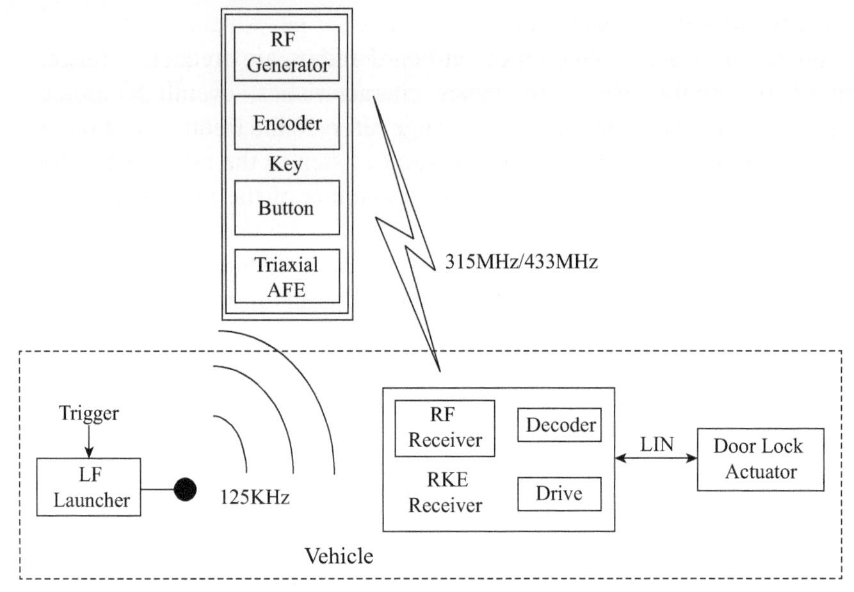

Figure 2.2 The block diagram of RKE system.

2.2 RKE/PKE Security

2.2.1 Overview of RKE

RKE system is also known as radio frequency (RF) key system. RKE performs exactly as a car key, which consists of a key fob transmitter and a receiver inside the vehicle and must work within a certain distance of the vehicle, usually 5–20 m. The block diagram of RKE system is shown in Figure 2.2.

The remote transmitter in the key fob and the receiver installed in the vehicle should work together. The former generally includes the radio transmission module and the encryption module. When the door unlocking or locking button, or trunk unlocking button on the remote transmitter is pressed, the microcontroller will wake up and send a series of 64-bit or 128-bit data stream to the RF transmitter of the key. After carrier modulation, the RF signal is transmitted to the space through the high-frequency oscillation circuit. The latter includes the radio receiving module and the decoding module, which receives the RF signal from the remote transmitter and sends the command to the actuator after decryption so as to obtain the desired actions such as unlocking or locking the door, or unlocking the trunk.

The transmitting frequency of the remote controller is usually set as 315 or 433 MHz, and the encoding modes are amplitude shift keying (ASK) and frequency shift keying (FSK) in general. In order to ensure a long effective remote control distance, the receiver needs higher sensitivity, wider receiving bandwidth, receiving resonant frequency close to transmitting frequency, receiving standing wave ratio close to 1, and antenna with good coordination with receiving LC network (also called inductor-capacitor network), etc. In order to ensure the security of RF communication, all the information sent by the remote controller is encrypted by some methods, which require the decryption of the received RF signal.

There are chips in the key fob of RKE system, and each chip has a fixed ID. Only when the ID of the key chip matches the ID of the engine can the vehicle start.

Take engine start-up as an example; when the user turns the key to start the vehicle, the in-vehicle base station sends low-frequency signal, which provides working energy for the transponder at the key fob, to start authentication. During the authentication process, the transponder in the key fob first sends its own ID number. The base station will send a series of random numbers and media access control (MAC) addresses once the ID number is verified by the base station chip, and the transponder will respond. The key fob will send a stream of data to start a session between the transmitter and the receiver. The data stream includes a preamble code, a command code, and a string of encrypted scroll codes. Among them, the scroll code type transformation is diverse. At least 50% of the number of bits of each encoding changes, and there is no rule for two encoding of the same operation.

The main functions of RKE transmitter are as follows. First, after a key-press, the encoding circuit is responsible for encrypting the corresponding key information and the information to be encrypted to form baseband signal. Second, the transmitting circuit modulates the baseband signal, by ASK in general, and emits it as RF signal. Third, there are two common schemes for transmitting circuit. One is based on IC, under which the circuit is simple and easy to implement, but the cost is high. The other is based on the separation device, which has the advantages of low cost, and the disadvantage is that the circuit is prone to be affected by the device error and difficult to match.

The RKE receiver filters, demodulates, decrypts, and decodes the transmitted data, restores the original one, and verifies the validity of it. If valid, the corresponding operations required by the user will be output.

The main functions of RKE receiver are as follows. First, the signal demodulation circuit demodulates the RF signal and restores it to baseband signal. Second, the baseband signal is decrypted according to the corresponding algorithm, and the key information is thus obtained. Third, the key information is sent to the master micro control unit (MCU) for verification and judgement to determine whether the corresponding action such as unlocking and locking should be performed.

Due to the characteristics of unidirectional transmission, the two biggest defects of RKE system are that the anti-interference ability is weak, and the transmitted RF signal is prone to be captured by other eavesdropping devices. For RKE system, the attacker will try to record the messages sent by the legitimate user when communicating with the vehicle. Later on, when the user leaves, the attacker replays the recorded messages to visit the vehicle or monitors RKE signal and sends the same frequency interference signal in combination with eavesdropping and interference when the user uses RKE system to lock the car. At this time, the RKE receiver may not receive the valid signal sent by the user, and the vehicle is still unlocked.

2.2.2 Overview of PKE

PKE system is designed to allow users to carry around an electronic key and get in the car without pressing it to perform remote unlocking. The system uses radio frequency identification (RFID) technology to automatically unlock and lock the door through a chip in the electronic key, which is a smart card, carried by the user. When the user tries to pull the door handle, the vehicle will send a query message to the smart card. If the authorized smart card is within the range of the vehicle and the valid verification code is answered, the vehicle can perform the corresponding action of unlocking or start-up. When the user leaves, the door will automatically lock and enter the anti-theft state.

PKE system uses the mode of low frequency trigger and high frequency authentication to complete two-way communication authentication. The authentication process is carried out through the encryption algorithm, which is up to 64 bits, between PKE in-vehicle transceiver and smart card, and between engine ECU and PKE. The vehicle can be started and operated only after the corresponding authentication. If there is no smart card, the electronic system of the vehicle cannot be started.

The main components of PKE system include in-vehicle transceiver, electronic key, antenna, etc. The in-vehicle transceiver is the critical part of the system, which is responsible for the communication with the electronic key and the interaction with the equipment. The electronic key, namely ID

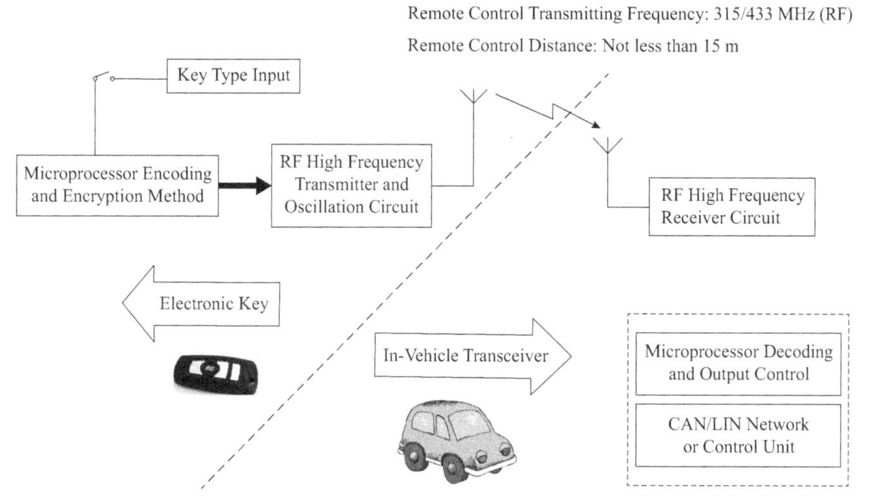

Figure 2.3 The block diagram of PKE system.

device, is carried around by the user and acts as the user's ID card, which is similar to the remote control of RKE system. Low-frequency (LF) antenna is the communication medium between the receiver and the electronic key, which receives and transmits RF signals.

PKE technology is also called the second generation technology of RKE. In short, it is actually the evolution from key pressing to no key and becomes safer. Figure 2.3 is the block diagram of PKE system.

The working process of PKE system can be divided into wake-up and verification.

- **Wake-up.** When the user with the electronic key appears in the detection range of the receiver and pulls the door handle, the electronic key will receive a low-frequency signal from the in-vehicle transceiver. If this signal matches the data stored in the key, it will be awakened. The design of wake-up mode can ensure that other irrelevant signals will not interfere with the operation of in-vehicle transceiver and can extend the service life of battery. In addition, the three-dimensional antenna used in PKE system can ensure that the key receives the wake-up signal without worrying about obstacles.
- **Verification.** After the key is awakened, it will analyse the password sent from the in-vehicle transceiver, calculate the corresponding data, and send it back after encryption. The host will analyse the data received

from the key and compare it with the data calculated by itself. If the two match successfully, the door will be unlocked. The whole process only takes a few milliseconds so that users will not feel any delay from touching the door handle to opening the door.

PKE system has the following advantages. PKE has absolute advantages over RKE in terms of function implementation and user experience. The user who uses RKE needs to press the remote control before unlocking and locking each time, while no operation is needed if he or she uses PKE. From the working principles of both, the two-way communication authentication mode in PKE is obviously more secure than RKE, which greatly reduces the possibility of being truncated and cracked. Especially, in the aspect of anti-jamming, RKE cannot work normally due to same frequency interference, while PKE will always let the door stay locked when it is interfered by the same frequency.

However, PKE also has shortcomings. The manufacturing cost of PKE system is higher than that of RKE because of its passive working principle, under which the power consumption of the system is relatively large, and the battery life of the electronic key is shorter than that of RKE.

2.2.3 Attack Technique Analysis of RKE/PKE

2.2.3.1 Security Technology of RKE/PKE System

RKE/PKE system uses various technologies to ensure the security of communication data, including scroll code technique based on Keeloq algorithm put forward by Microship, authentication technology based on Hitag2 encryption algorithm proposed by NXP, and security technology based on AES encryption algorithm by TI.

- **Security strategy based on Keeloq algorithm.**
  Keeloq algorithm was a block encryption algorithm, originally designed by Willem Smith of South Africa. It was purchased by Microchip in 1995 and a series of special codec chips were launched accordingly. At present, there are still a lot of vehicles that apply this algorithm in RKE/PKE systems. The core idea of Keeloq algorithm is to encrypt 32-bit plaintext with 64-bit key so as to get 32-bit ciphertext. Even if only one bit of data in plaintext changes, more than 50% of the ciphertext obtained by Keeloq algorithm will change.
  Most of the RKE systems use Keeloq hopping code as the input variable of Keeloq encryption. The transmitter and receiver separately integrate

an encoder and a decoder and share a key and a set of fixed identification code to distinguish different RKE systems. At the same time, there is a group of 32-bit synchronous code, adding one bit after each successful signal transmission to distinguish the information sent each time. Finally, it is a 4-bit function key information. The encoder performs Keeloq encryption by inputting 32-bit synchronous code and 64-bit key information stored in advance and obtains the hopping code in the end.
The PKE system uses the Keeloq IFF two-way verification in the form of question and answer, in which the vehicle and the key share an encryption key (fixed password). For the application of question-answering verification mode, it generally adopts two communication frequency bands. LF frequency band (315 kHz) is used for short-distance communication, which is mainly used to detect whether the key is inside or outside the vehicle. Ultra high frequency (UHF) frequency band (315 MHz/433 MHz) is used for long-distance communication, mainly for sending verification messages. Its working mode can be divided into normal working mode and backup working mode. In normal working mode, PKE system can be powered by battery, while in backup mode, it does not need battery support.

- **Security strategy based on Hitag2 algorithm.**
Hitag2 algorithm was first used in the keyless entry system introduced by NXP company and is the mainstream used in PKE system to ensure vehicle safety currently. Developed from Crypto1, Hitag2 algorithm has a lot of improvement. The key is not directly entered into linear feedback shift register (LFSR), but processed separately. Now, Hitag2 card and some other RFID chips have been widely used in security solutions of automotive anti-theft system and other systems.
The Hitag2 encryption unit is mainly composed of a 48-bit LSPR and a nonlinear filter function. In each clock cycle, the 20-bit output of LFSR is calculated by nonlinear filtering function to generate a password, and then the data in LFSR shifts one bit to the left. Meanwhile, the 48th bit of LFSR is calculated by the password, device ID, and key generated in the previous step. The security of Hitag2 has been focused since 2007, and many attacks have been published aiming at it. These attacks mainly depend on the following conditions: the 48-bit key length of Hitag2 algorithm, the weakness of filter function, and the feature that internal state of a given transponder in each session in the first 48 cycles remains unchanged.

- **Security strategy based on AES.**
  AES is an iterative, symmetric key block cipher. It can use 128-, 192-, and 256-bit length keys, encrypting and decrypting data with a 128-bit block. The encryption process of AES is realized by iterating the input plaintext and the key by the round function through N_r rounds. The operations named SubBytes, ShiftRows, MixColumns, and AddRoundKey are performed in the first $N_r - 1$ round, while MixColumns is removed in the last round. The decryption process is opposite. The input ciphertext and the key are iterated by the round function. In the first $N_r - 1$ round, the operations called InvShiftRows, InvSubBytes, AddRoundKey, and InvMixColumns are carried out in turn, while the last round does not need InvMixColumns.

2.2.3.2 Summary of Common Attacks of RKE/PKE System

At present, the main attack methods against RKE/PKE system include brute-force scanning attack, replay attack, relay attack, forward prediction code attack, dictionary prediction attack, and cryptanalysis attack. The attack methods are different against RKE system and PKE system, which are briefly introduced below.

- **Brute-force scanning attack.**
  The brute-force scanning attack continuously sends different attack codes to the RKE system, aiming at its scrolling code technology. The attacker will keep trying until the transmitted codes match.
  But things are slightly different as for PKE system. The attacker will try to get access to the vehicle by pulling the door handle several times. The attacker will return a fixed code to the vehicle each time, the main purpose of which is to let the vehicle send back the corresponding query code to match the fixed code sent on.
  Brute-force scanning attack is very simple from the attacker's point of view because no technical information about the system is required. The success probability of it depends on the number of bits randomly interrogated by the vehicle, the encryption algorithm of random query, and the number of experiments conducted by the attacker.
  As one of the most basic attack methods, the main methods against brute-force scanning attack include increasing the bit length of key to increase the time required by an attack and adding random redundant information in plaintext or ciphertext, etc.

- **Replay attack.**
  Replay attack is also known as signal recording attack. The attacker will try to record the message sent by the legitimate user when communicating with the vehicle. When the user leaves, the attacker attempts to access the vehicle by replaying the recorded message. The attack can be extended to combine eavesdropping and jamming. For example, when the user uses RKE system to lock the car, it monitors the transmitted signal and sends the co-frequency interference signal. In this case, the RKE receiver may not receive the valid signal sent by the user, and the vehicle will still be in the unlocking state while the attacker has temporarily valid code information.
  This type of attack can only be effective on the system that the information verified by each transmission is invariant. It will not work when the scroll code technology or challenge-response verification method is used. For example, if the receiver only receives a valid signal, while the RKE system will generally send at least 4–5 frames of the same signal after pressing the button once, then the signal monitored by the eavesdropper will be regarded as the previous signal and judged as invalid.
- **Relay attack.**
  Relay attack requires two attackers to cooperate with each other and use an appropriate electronic equipment to establish a relay between the vehicle and the electronic key. The first attacker stands next to the vehicle and pulls the door handle to receive the signal sent by the vehicle and then amplifies and sends it to the second attacker who stands near the owner. The second attacker sends the signal to the owner after receiving it; then the key will respond. The second attacker receives the key's response and sends it back to the first one, who then sends it to the vehicle, and the vehicle will open the door. It will succeed as long as an appropriate electronic equipment is chosen.
- **Forward prediction code attack.**
  Forward prediction code attack means the attacker tries to predict what code will be sent by the vehicle in the next time by observing those sent in the previous times. In PKE system, the random code of multiple queries can be easily obtained by pulling the door handle several times. If the attacker can predict the next random query code via some method, he can go near the electronic key to generate a predicted random access code and then record the corresponding key response. After that, the attacker can return to the vehicle and pull the door handle to trigger the

system. By replaying the message recorded to the vehicle, the unlocking and disarming can be completed.

- **Dictionary prediction attack.**
  Dictionary prediction attack is also aimed at PKE system. The attacker will record the query and answer of vehicles to build a dictionary, in which each entry contains a valid query–answer pair. The attacker can simply send the pre-edited random code near the electronic key, capture the key response corresponding to each random code, and store the pair in the dictionary accordingly. Once the attacker has successfully established the dictionary, he can continuously pull the door handle until the vehicle generates a random code which is included in the dictionary. The corresponding key response can be found and transmitted, and, thus, the vehicle can be unlocked and disarmed.
- **Congestion attack.**
  Another way of attacking keyless entry system is to perform signal congestion attack with radio jamming equipment. If the door is fully manually controlled, it will only be locked by key-press, and the vehicle will stay open when the door locking signal is congested.
- **Other effective attacks.**
  Other effective attacks include using social engineering methods to obtain internal or critical information from vehicle manufacturers, key solution providers, automotive repair shops, parts suppliers, etc., to copy or crack the keys.

2.3 TPMS Security

2.3.1 Overview of TPMS

TPMS is mainly composed of controller, sensor and instrument panel. Figure 2.4 shows a TPMS.

The tire pressure monitoring sensor integrates pressure sensor, temperature sensor, acceleration sensor, and battery voltage sensor, which can detect those information in real time. For example, when the acceleration sensor detects that the rotation acceleration exceeds a certain threshold value, the collected information is encoded and sent to the tire pressure monitoring controller through RF signal, which can decode and get the current tire pressure and temperature status. When the tire pressure is not in the preset range or the tire temperature or air leakage speed exceeds the predetermined threshold, TPMS will send out the corresponding alarm information, which

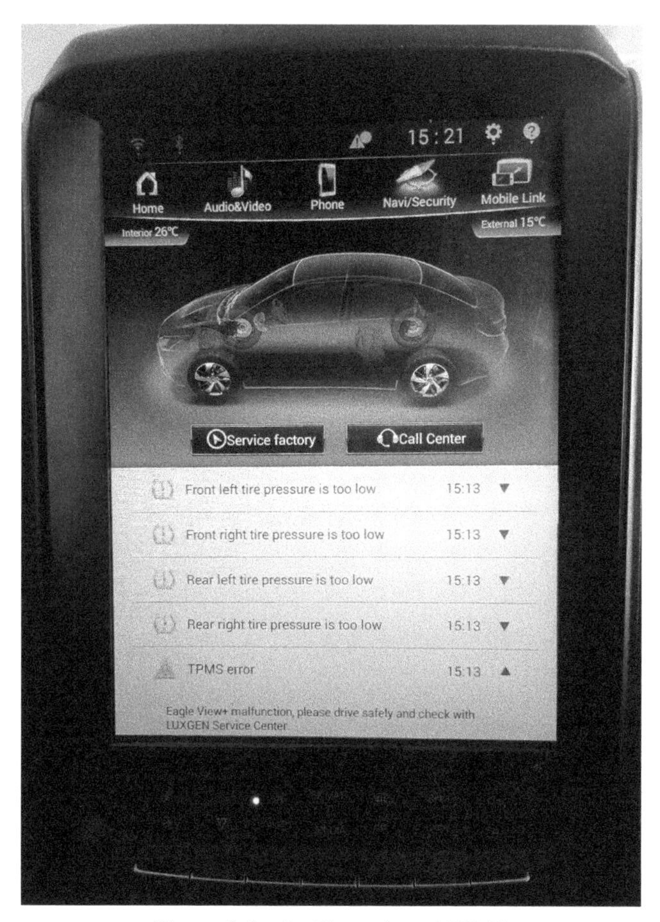

Figure 2.4 An illustration of TPMS.

will be transmitted to the instrument panel through controller area network (CAN) bus network.

At present, TPMS can be mainly divided into three categories, that is, direct TPMS, indirect TPMS, and hybrid TPMS.

- **Direct TPMS.**
  Direct TPMS is to install a pressure sensor in each tire to directly measure its pressure, which will be transmitted to the central receiver through a wireless transmitter, and the pressure data will be shown on the vehicle display panel. When the tire leaks or the pressure is low, the system will raise alarm automatically.

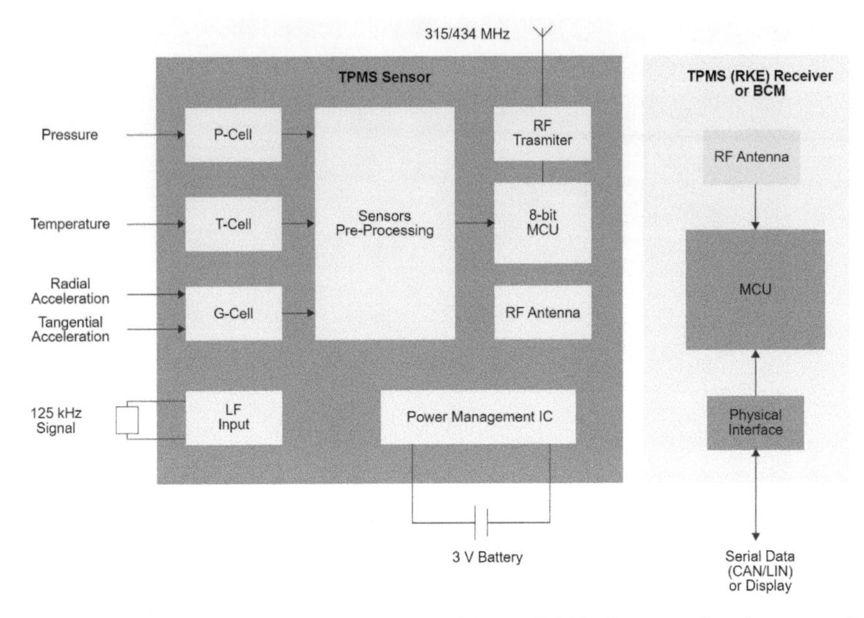

Figure 2.5 Specifications of a TPMS provided by NXP. This figure needs to be removed.

Direct TPMS has to bear high cost while enjoying high precision. It can be divided into passive and active power supply modes.

1. Passive TPMS is also called battery-free TPMS. It uses a transceiver to replace the general receiver. The transponder installed in the tire receives the signal from the transceiver and uses the energy of the signal to emit a feedback signal back to it. This makes the pressure and temperature sensors inside the tire complete data transmission without the battery, thus solving the problem of limited lifespan of battery. Passive TPMS does not need batteries, but it needs to integrate the transponder into the tire module, which has not formed a unified standard yet; so it is not the mainstream.
2. Active TPMS uses the sensors inside each tire that need power supply to directly measure the tire pressure and temperature and sends the data to the receiver of the host module through the RF transmitter. The host module completes data analysis, processing, and display. At present, the developed module of active TPMS can be applied to the tires produced by various tire manufacturers. However, in active TPMS, the limited battery life is still a major problem.

- **Indirect TPMS.**
  Indirect TPMS uses non-pressure sensor to measure the relevant data and calculates the tire pressure by tire mechanical model or comparing the pressure difference between tires. It has the advantages of simple structure, low cost, and strong durability compared with direct one. There are three kinds of monitoring methods in aspects of speed, frequency, and magnetic sensitivity.
  1. **Speed monitoring method.** It uses the wheel speed sensor of ABS system to compare the speed difference between tires to monitor tire pressure. When the pressure of a tire decreases, the rolling radius of it will become smaller, resulting in the faster speed of the wheel than any other; so the tire pressure can be monitored.
  2. **Frequency monitoring method.** While on the road, the spring index of tire often varies with the tire pressure. By processing the waveform generated by ABS wheel sensor installed on the wheels, the resonance frequency of tire can be calculated; thus, the spring constant of tire can be obtained. According to the proportional relationship between tire and spring constant, tire pressure can be obtained.
  3. **Magnetic sensitivity monitoring method.** The tire pressure sensor is installed on the wheel rim, and the hall device is installed on the bracket fixed with the suspension strut or the wheel brake backing plate. When the car is running, the change of tire pressure will cause the magnetic field direction of the magnetic component in the pressure sensor to change, which indirectly changes the magnetic induction intensity of the magnetic sensor of the hall device, and the output signal of it also changes. This realizes the non-contact transmission of inflation pressure signal from tire to vehicle body. The electronic control unit (ECU) is composed of single-chip microcomputer and peripheral interface, the former of which samples the output signal of the hall device after conditioning and sends the data to the memory. After calculation, analysis, and comparison, tire pressure can be obtained and displayed on the panel or makes sound and light alarm when the pressure is abnormal.
- **Hybrid TPMS.**
  It takes into account both cost and accuracy of tire pressure monitoring. Tire pressure sensors and an RF transceiver are installed on the two

opposite wheels to increase the measurement accuracy. The current tire pressure monitoring alarm can be summarized as navigation/DVD upgrade type, cigarette lighter type, portable type, independent type, ceiling type, and external type. Each tire is equipped with a highly sensitive sensor to monitor the pressure and temperature of tire in real time, and the data signal is transmitted to the terminal receiving system in the vehicle. In case of air leakage, high or low tire pressure, high temperature, and other abnormal conditions, the terminal will automatically alarm and display the tire status to ensure the safety of driving.

2.3.2 Attack Technique Analysis of TPMS

TPMS presents potential risks as the use of it is not voluntary and it is hard to deactivate. A dashboard warning caused by tire pressure spoofing will likely cause the driver to pull over and inspect the tire.

- **Reverse engineering.** TPMS communications are based on standard modulation schemes and simple protocols, which do not rely on cryptographic mechanisms. Therefore, the communication in it can be reverse-engineered. To elaborate, the modulation schemes, encoding schemes, and message formats are what one needs to know. Also, the activation and reporting methodologies to properly decode or spoof sensor messages are necessary. This information requires reverse-engineering by an adversary apart from access to an insider. By collecting the transmissions from each tire pressure monitoring sensor, information between the activation signals and the sensor responses will be timely extracted. Once the sensor bursts are collected, the signal analysis can be conducted to determine the coarse physical layer characteristics at first, then the modulation and encoding schemes can be identified as well, and the message format will be mapped out finally.
- **Eavesdropping.** Although tire pressure data does not need strong confidentiality, TPMS protocol contains an identifier that can be used to track the location of equipment. In fact, the probability that a transmission can be observed by a stationary receiver depends not only on the communication range but also on the messaging frequency and speed of the vehicle under observation. These factors affect the occurrence of transmissions in communication range.

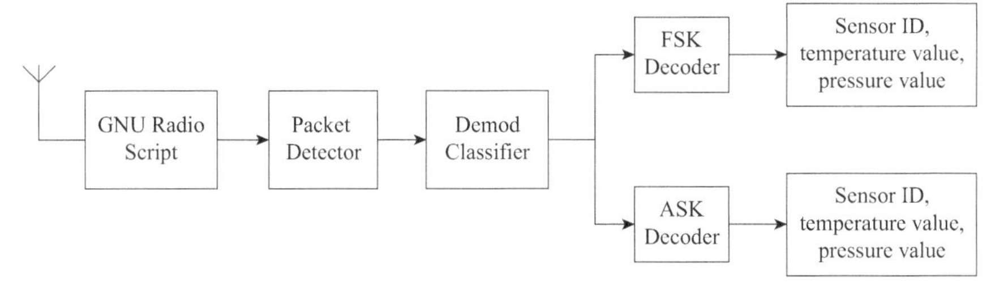

Figure 2.6 The block diagram of a real-time eavesdropping system.

The transmission power of the pressure sensor is relatively small to extend the sensor battery life and reduce cross-interference. In addition, National Highway Transportation Safety Administration (NHTSA) requires tire pressure sensors to transmit data only once every 60–90 seconds. The low transmission power, low data reporting rate, and high driving speed of vehicles affect the feasibility of eavesdropping.
Figure 2.6 shows a block diagram of a real-time eavesdropping system. The GNU radio script is to sample channels at a certain rate, where the recorded data is piped to a packet detector. Once the packet detector identifies high energy in the channel, it can extract the complete packet and pass the corresponding data to the decoder to obtain the pressure, temperature, and the sensor ID. A modulation classifier is applied in the eavesdropping system to recognize the modulation scheme and choose the corresponding decoder. If the decoding process is successful, the sensor ID will be output to the screen, and the raw packet signal with the time stamp will be stored for later analysis. In this way, diverse data can be captured from multiple different TPMS.

- **Packet spoofing.** Compared with eavesdropping signals, sending forged data packets poses a greater risk.
  Eavesdroppers can monitor and decode the messages of tire pressure sensors in real time. On the basis of eavesdropping system mentioned above, the packet generator can take two sets of parameters, namely sensor type and sensor ID from the eavesdropper, and pressure, temperature, and status flags from users, to generate a properly formulated message. It then modulates the message at baseband while inserting the proper preamble. The forged sensor packets are up-converted and transmitted either continuously or just once at the desired frequency. Once the sensor ID and sensor type are captured, the forged message

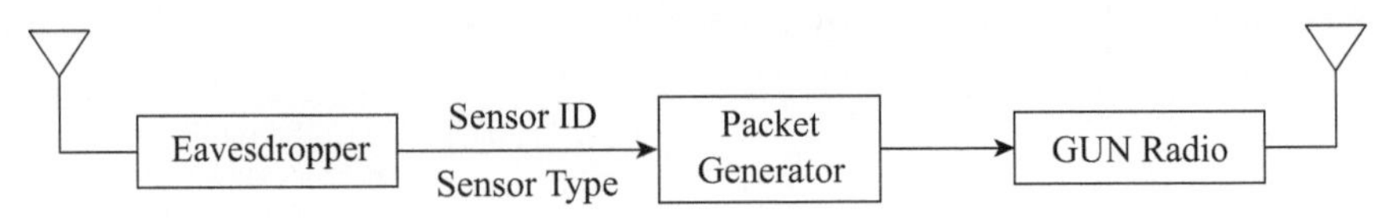

Figure 2.7 A demonstration of packet spoofing based on the eavesdropping system.

can be created and repeatedly transmitted at a pre-defined period. Figure 2.7 is a demonstration of packet spoofing.

2.4 On-Board Wi-Fi Security

2.4.1 Overview of On-Board Wi-Fi

On-board Wi-Fi is a wireless Internet service launched for public transportation such as buses, coaches, and private cars. Wi-Fi terminals get access to the Internet to obtain information, entertainment, and mobile office business in a wireless manner. On-board Wi-Fi devices refer to wireless devices that provide Wi-Fi hotspots via 3G/4G/5G, wireless RF, and other technologies loaded on vehicles. Figure 2.8 shows the in-vehicle Wi-Fi.

IVI can be equipped with a Wi-Fi function module by configuring a Wi-Fi chip. The Wi-Fi module can be connected to a wireless hotspot of a mobile phone or a Wi-Fi access point (AP) base station so as to realize part of the Internet of vehicles, improve the audio and video entertainment system in the vehicle, and increase the Internet function for users, which also has become a path for hackers to attack vehicles. Figure 2.9 is an example of Wi-Fi attack path.

The transmission distance of Wi-Fi is getting farther and farther, and there are more and more base stations. Seamless handover becomes increasingly easier; so it may become the main body of the Internet of vehicles in the future, and the probability of being attacked is also very large.

Generally speaking, an attacker can control the IVI system by cracking the Wi-Fi authentication encryption mode to further control the in-vehicle entertainment equipment, such as changing the radio band, starting and closing the navigation system, adjusting the level of volume, etc. Additionally, the attacker may penetrate the CAN bus to control other ECU modules, resulting in serious consequences.

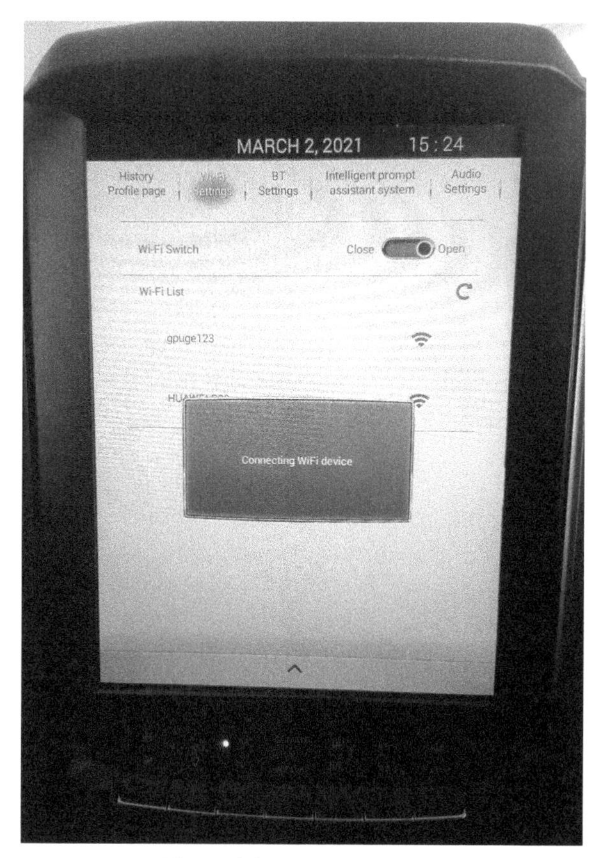

Figure 2.8 In-vehicle Wi-Fi.

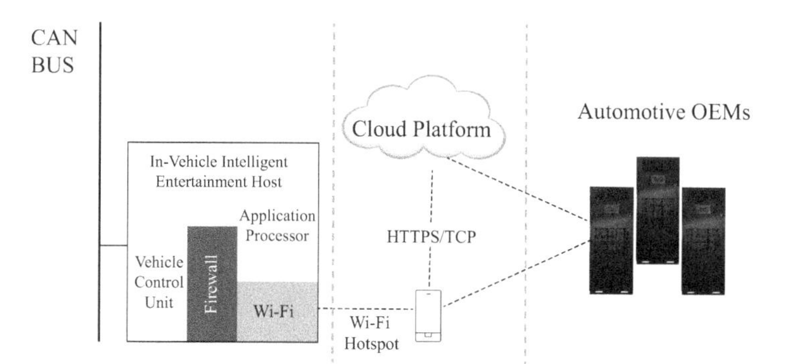

Figure 2.9 An illustration of Wi-Fi attack.

2.4.2 Attack Technique Analysis of On-Board Wi-Fi

Due to the particularity of Wi-Fi in terms of mobile devices and transmission media, some attacks are easy to implement. Because there are multiple APs in the early days, it is impossible to determine whether users have obtained security certification when accessing the network, which brings the security problem of the WLAN.

Having faced with this situation, IEEE and other organizations, such as the Wi-Fi Alliance, have published many security standards to deal with different security vulnerabilities. This chapter will introduce several major WLAN security protocols, that is, wired equivalent privacy (WEP), WPA, WPA2, and WPA3, and briefly introduce the corresponding attack methods against these protocols.

2.4.2.1 WEP Protocol

WEP is the earliest security protocol for WLANs which is based on link layer protection. It enables WLANs to have the same security performance as traditional wired networks. WEP uses the RC4 encryption algorithm to protect the confidentiality of data. The AP and the terminal use the same key to authenticate, and the cyclic redundancy check (CRC) value is used to ensure the integrity of data at the same time. However, this protocol was found with security flaws shortly after coming out.

The encryption process of WEP protocol is shown in Figure 2.10. The integrity check value (ICV) is calculated using CRC-32 algorithm and is appended to the end of the plaintext. The 24-bit initialization vector (IV) and the 40-bit key K are mixed into the key stream seed, and the pseudo-random sequence generator (PRSG) is then used to generate the key stream. The plaintext and the key stream are XOR-added to obtain the ciphertext.

- **Major security threats of WEP.**
  The threats faced by the WEP protocol are mainly divided into two categories according to the concern with RC4 algorithm.
  1. One is that related to RC4 algorithm. In the WEP encryption process, the shared key and IV form the initial key stream seed, which generates the encryption key through RC4, and the ciphertext is obtained through the encryption key and the plaintext by XOR operation. Therefore, by intercepting two more ciphertexts, the other ciphertext can be inferred when a piece of plaintext is known.
  2. The other is the threats that have no concern with the RC4 algorithm. On the one hand, although CRC is used to verify the integrity

of data; it cannot provide a complete security check code. Instead, the CRC check determines whether the transmitted data has been tampered with by the ICV value at the end of the encrypted data packet. Therefore, the attacker can easily fabricate a data packet with the same ICV value, leading to a failure of verification of data integrity.

On the other hand, the WEP protocol does not have a complete key management system. It completes the authentication of the entire network only through a simple shared key, which will lead to the leakage of user data in the whole network once the key is leaked. Moreover, WEP adopts a one-way authentication process, which is vulnerable to MITM attacks.

- **Attack methods against WEP.**
  Attack methods against WEP protocol are also divided into two categories accordingly. Most attack methods have been implemented by tools such as Aircrack-ng.

  1. Attacks related to RC4 are mainly aimed at RC4 in WEP or WEP-like environments. The goal of the attacks is to obtain a key instead of just a pseudo-random sequence. The main attack methods are FMS attack, KoreK attack, Mantin second round attack, Klein first round attack, etc.
  2. There are four main types of attacks unrelated to RC4.
     - **Packet injection attack.** The attacker captures the packets of the WEP network and replays them after a period of time. The injection attacks are divided into normal injection and address resolution protocol (ARP) injection.

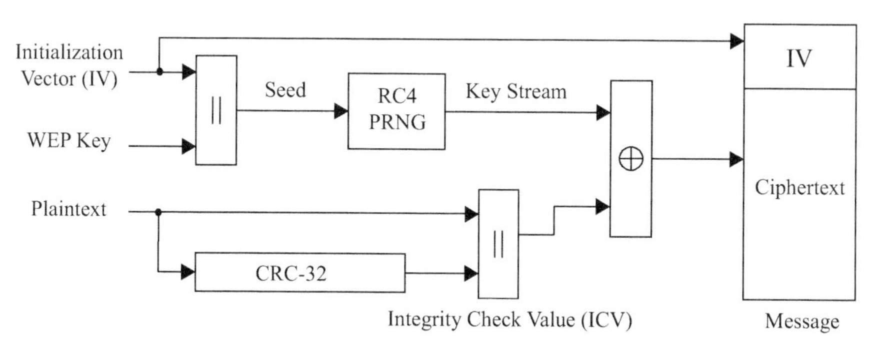

Figure 2.10 The encryption process of WEP protocol.

- **Authentication attack.** The attacker captures the authentication data packet exchanged between the station (STA) and the control AP to construct a new legal authentication.
- **Chopchop attack.** This attack uses WEP to attack the weaknesses in the CRC-32 check process. The attacker obtains a ciphertext of length L and can crack the last m-length plaintext of the ciphertext and a pseudo-random sequence for encryption, which requires an average of $128m$ basic attack operations. The reverse version of this attack method is the Arbaugh attack.
- **Fragmentation attack.** The attack utilizes the weaknesses that 802.11 standard allows data segmentation. The attacker obtains a pseudo-random sequence for encryption of length m. Through data segmentation, a data payload of length 16 (m-4) can be sent, and a pseudo-random sequence for encryption of length ($16m$-60) can be thus obtained.

2.4.2.2 WPA/WPA2 Protocol

WEP was found to be insecure soon after its release. Wi-Fi Alliance developed a security mechanism called Wi-Fi protected access (WPA) on the basis of the 802.11i draft, which uses the temporal key integrity protocol (TKIP) mechanism and the same RC4 encryption algorithm used in WEP. So, there is no need to modify the hardware of original wireless devices. WPA addresses the problems existing in WEP, such as excessively simple key management and no effective protection for message integrity, and the security of network is improved through software upgrades. It strengthens the user authentication function provided by WEP; and includes the support for 802.1x and extensible authentication protocol (EAP). pre-shared key (PSK), and remote authentication dial in user service (RADIUS) are used for authentication.

WPA encrypts with constantly changing keys, which makes it more difficult to invade wireless networks than WEP. The PSK method is used in small WLANs and home networks, which only requires a key to be pre-entered in each WLAN node. As long as the keys match, users can gain access to the WLAN. The more long and complex the PSK password is, the more difficult and time-consuming it is for hackers to crack, therefore improving the security of wireless networks. In this case, WPA uses TKIP to establish a dynamic key and mutual authentication mechanism for encryption. The security function of TKIP makes up for the deficiencies of WEP and provides

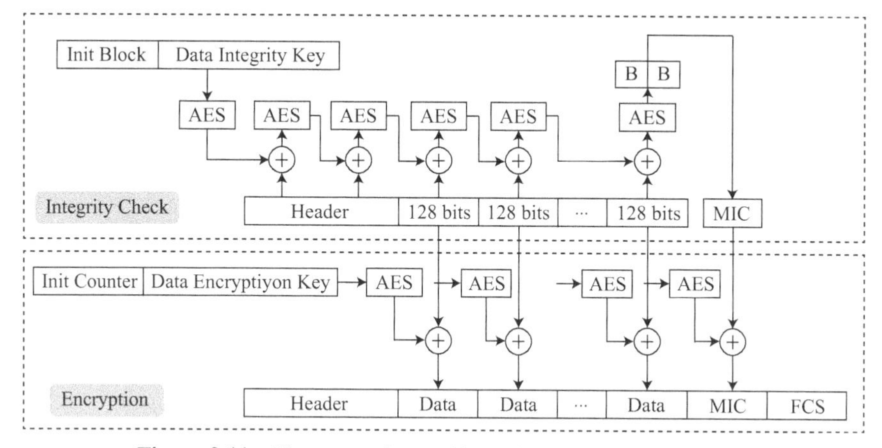

Figure 2.11 The encryption and integrity of the WPA2 protocol.

a high level of security for small WLANs and home users. On the other hand, the RADIUS server authentication method is used in commercial and enterprise-level WLANs. Message exchange during the process is verified via the EAP, which carries the credentials of users for authentication, such as the user name and password. 802.1x is used to verify the user's identity through the RADIUS server and provide enterprise-level security authentication for WLANs.

WPA2 is the second generation of WPA, which implements all mandatory features from 802.11i. It provides a significant security improvement over WEP and WPA. The most important architectural change is the data encryption algorithm. In WPA2, the counter mode with CCMP (cipher block chaining message authentication code protocol) uses an AES (advanced encryption standard).

Both integrity check and encryption are based on AES algorithm. Figure 2.11 is a block-based algorithm, operating on 128-bit chunks. Input message is divided into 128-bit parts and used together with previous results as an input to the AES algorithm. The whole message, including header, is taken into consideration for integrity check, while only payload is encrypted. Header must be transmitted as plaintext in order to enable frame detection and decoding.

- **WPA/WPA2 four-way handshake.** WPA and WPA2 are almost the same at the technical level. The main difference is that WPA2 requires the support for more secure CCMP. Both WPA and WPA2 use the four-way handshake defined in 802.11i. Pairwise transient key (PTK) is

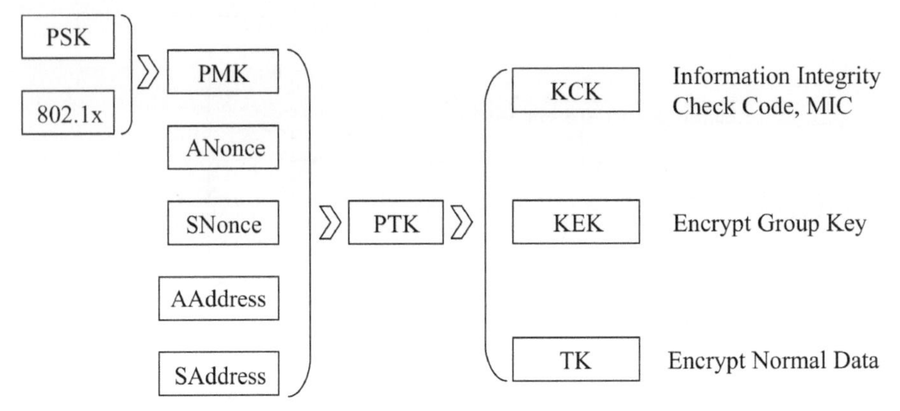

Figure 2.12 The relationship of keys and parameters in four-way handshake.

calculated by pairwise master key (PMK), AP random number ANonce, STA random number SNonce, and MAC addresses of both parties. Among them, PMK is calculated and generated by information known to both parties, such as login password. The temporal key (TK) used for subsequent normal data encryption is derived from PTK. Figure 2.12 illustrates the relationship of keys and parameters in four-way handshake.

WPA/WPA2 uses a four-way handshake to generate the required PTK from PMK through a series of interactions. Figure 2.13 shows the four-way handshake process, which can be summarized as follows.

1. AP sends ANonce to the terminal. After receiving this message, the terminal has all the elements for generating PTK because the MAC address of the AP is also included in the message.
2. The terminal calculates the PTK and sends SNonce and its own MAC address to the AP. At the same time, the message integrity code (MIC) is calculated by the key confirmation key (KCK) in the PTK and sent to the AP.
3. AP receives packet 2 and calculates PTK, and then calculates the MIC to compare with the MIC in packet 2. If the same, then send an authenticated packet, which also contains MIC.
4. After receiving the message 3, the terminal calculates whether the MIC is the same as the MIC in the message 3. If the same, it means that the PTK has been installed and the following data can be encrypted.

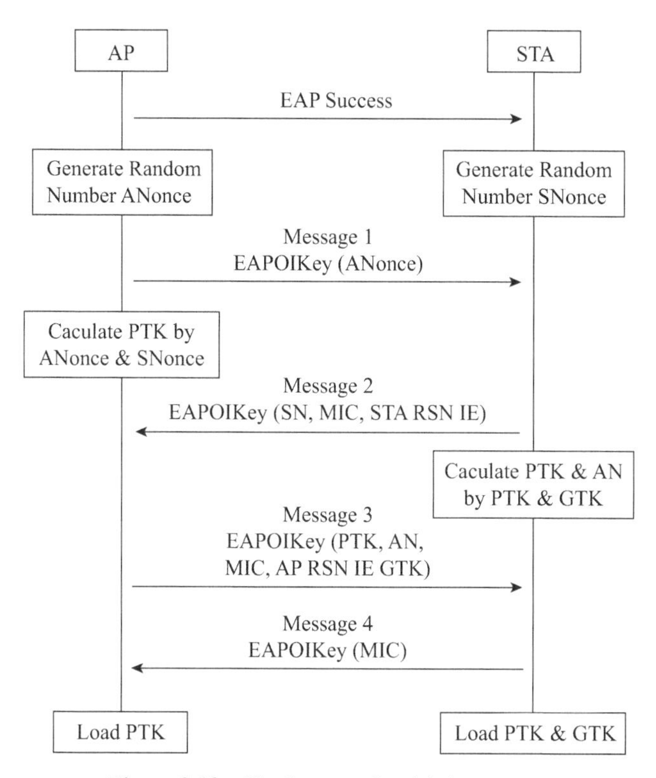

Figure 2.13 The four-way handshake process.

- **Main attack methods against WPA/WPA2.** The WPA/WPA2 protocol was originally designed to fix some security problems exposed by the WEP protocol. Although it is resistant to replay attacks, weak key attacks, and statistical attacks to a large extent, with the emergence of new cracking methods, there have been some studies that found the security weakness of WPA/WPA2 protocol.

  1. **Key attack.** The WPA protocol uses TKIP encryption to solve the data security, but TKIP is essentially extended on the WEP protocol. It still uses the RC4 algorithm internally and cannot fundamentally solve the weakness of the RC4 algorithm itself. Therefore, although the TKIP mechanism strengthens the transmission security to a certain extent, the key can still be cracked by grabbing the data packet. The attacker can design a key dictionary, calculate the PTK value by assuming that a certain key is correct,

and then compare this PTK with the captured PTK value. If the same, it proves that it is the key that the AP authenticates with the terminal.

2. **Man-in-the-middle attack.** The authentication mechanism of the WPA/WPA2 protocol can be divided into two categories, one is the personal authentication based on PSK and the other is the server authentication based on 802.1x and EAP. Although the server authentication mode is a two-way authentication mode, it is not between the terminal and the access point but the access point and the authentication server. Therefore, in both modes, the attacker can intercept the message and then forge a message that can be accepted by the receiver, thereby causing a MITM attack.
3. **Denial of service attack.** This attack method is to stop the target machine from providing network services, which is a very serious method of malicious attacks on both wired and wireless networks. In WLANs, DoS attacks are mainly divided into three types, i.e., physical layer attacks, MAC layer attacks, and protocol layer attacks. The physical layer attack is mainly to destroy the normal channel by transmitting jamming RF signals. The MAC layer attack is mainly to maliciously occupy the communication channel for a long time. In the four-way handshake negotiation process, because the management frame and the control frame do not use any protective measures, the attacker can easily forge these management frames and control frames to launch the authentication process, which prevents normal users from connecting. The protocol layer attacks mainly include DoS attacks against EAP.TLS authentication and DoS attacks against the four-way handshake protocol.
4. **KRACK/Key reinstallation attack.** The attack targets the four-way handshake used to establish a nonce, which is a kind of "shared secret", in the WPA2 protocol. The standard for WPA2 anticipates occasional Wi-Fi disconnections and allows reconnection using the same value for the third handshake for quick reconnection and continuity. Because the standard does not require a different key to be used in this type of reconnection, which could be needed at any time, a replay attack is possible.
5. **Brute-force/dictionary attack.** This relies on capturing a WPA handshake and then using a wordlist or brute-force to try and crack the password. Depending on the password strength, such as length,

charset, it can be difficult or impossible to break it in a reasonable amount of time.

2.4.2.3 WPA3 Protocol

A fundamental weakness of WPA2 is that it lets hackers deploy a socalled offline dictionary attack to guess the password. An attacker can take as many shots as they want at guessing the credentials without being on the same network, cycling through the entire dictionary in relatively short order. In January 2018, the Wi-Fi Alliance announced WPA3 as a replacement to WPA2.

WPA3 will protect against dictionary attacks by implementing a new key exchange protocol. The aim of WPA3 is to simplify security, enable robust authentication, and increase cryptographic strength. WPA2 used an imperfect four-way handshake between clients and access points to enable encrypted connections; it is what was behind the notorious KRACK vulnerability that impacted basically every connected device. WPA3 will ditch that in favour of the more secure simultaneous authentication of equals (SAE) handshake, which is depicted in Figure 2.14.

SAE handshake is based on key exchange and is resistant to offline dictionary attacks, where previously recorded data can be used to guess the

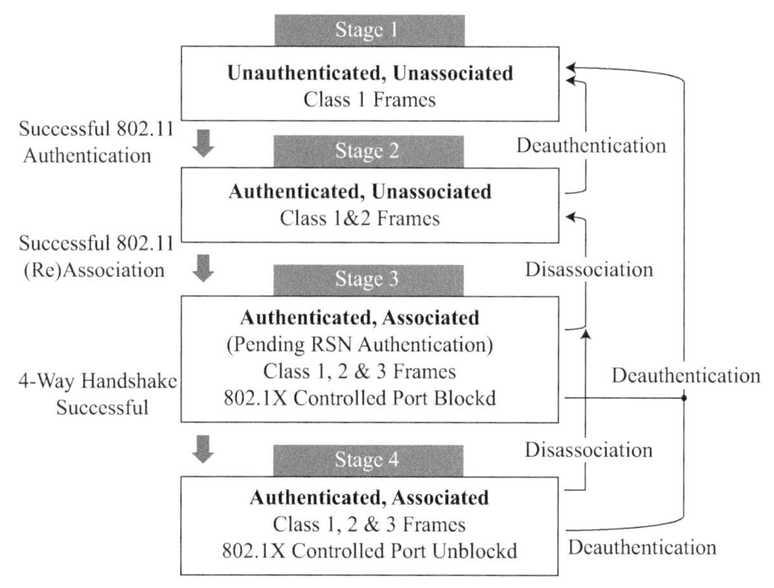

Figure 2.14 The simultaneous authentication of equals handshake in WPA3.

password. This mechanism allows to protect the data even if the password is weak, meaning is short or using popular phrases. However, widespread adoption of WPA3 will not happen overnight due to the hardware limitation. Table 2.1 compares WEP, WPA, WPA2, and WPA3 protocols.

2.5 On-Board Bluetooth Security

2.5.1 Overview of On-Board Bluetooth

As a narrow bandwidth transmission technology, Bluetooth is open, compatible, and portable. The specification of Bluetooth wireless communication technology is completely open and shared for different manufactures in need to apply. Interoperability and data sharing can be realized between Bluetooth products of different companies, and it can be applied in many occasions.

Nowadays, the applications of Bluetooth communication in vehicles mainly include automotive Bluetooth key, Bluetooth hands-free communication, Bluetooth rearview mirror, on-board Bluetooth entertainment system, on-board Bluetooth self-diagnosis technology, on-board Bluetooth anti-theft system, automotive steering wheel control system, and so on.

- **Automotive Bluetooth key.** The core of it is the key control software. Car users can intelligently control the car Bluetooth terminal with one click on the Bluetooth Key App to realize the functions of unlocking and locking the vehicle, and opening the trunk.
- **Bluetooth hands-free communication.** It is to let on-board phone access the user's mobile phone SIM card via Bluetooth and identify the information including mobile phone number, service provider, user ID, contact, etc., and can automatically login the network of phone operator, realizing the wireless connection between the user's mobile phone and on-board phone.
- **Bluetooth rearview mirror.** It is a new type of on-board phone that connects to the user's mobile phone via Bluetooth. The rearview mirror can display the number of incoming calls in the mirror and integrate the function of hands-free call.
- **On-board Bluetooth entertainment system.** Nowadays, all-in-one navigation machines are widely received by the public. On the basis of on-board GPS navigation, it has added Bluetooth phone function, which not only can answer and make calls but also realize the memory communication with the smart phone, showing the pictures, audio, and video files in the on-board machine.

Table 2.1 The Comparison Among WEP, WPA, WPA2, and WPA3.

	WEP	WPA	WPA2	WPA3
Brief Description	Ensure wired-like privacy in wireless	Based on 802.11i without requirement for new hardware	All mandatory 802.11i features and a new hardware	Announced by Wi-Fi Alliance
Encryption	RC4	TKIP + RC4	CCMP/AES	GCMP-256
Authentication	WEP-Open WEP-Shared	WPA-PSK WPA-Enterprise	WPA2-Personal WPA2-Enterprise	WPA3-Personal WPA3-Enterprise
Data Integrity	CRC-32	MIC algorithm	Cipher block chaining message authentication code (based on AES)	256-bit broadcast/multicast integrity protocol Galois message authentication code (BIP-GMAC-256)
Key Management	None	four-way handshake	four-way handshake	Elliptic curve Diffie–Hellman (ECDH) exchange and Elliptic curve digital signature algorithm (ECDSA)

- **On-board Bluetooth self-diagnosis technology.** It is sending the vehicle self-diagnosis function via Bluetooth to intelligent devices. Through the intelligent device, users can quickly receive the fault code in vehicle and its corresponding meaning, allowing them to evaluate the performance and state of the vehicle and ensure the safety.
- **On-board Bluetooth anti-theft system.** The working principle of the existing Bluetooth anti-theft system of vehicle door is to match the mobile phone Bluetooth with the on-board one and lock or unlock the door through the mobile phone app, while that of engine anti-theft system is to control the start circuit of engine through Bluetooth – when the mobile phone Bluetooth signal cannot be found by on-board Bluetooth or is found to be illegal, the engine cannot start. Some can also remotely start the vehicle and air conditioning.
- **Automotive steering wheel control system.** It uses Bluetooth to realize the electronic optimization design of the steering wheel panel switch, overcoming the disadvantage of looking for the switch during driving, and make most of the operations realized on the steering wheel.

2.5.2 Attack Technique Analysis of On-Board Bluetooth

The core of Bluetooth technology is to realize wireless connection. Compared with the traditional wired connection, the wireless interface can be better used to realize device interconnection through Bluetooth, and the portability is very strong. In practice, Bluetooth is aimed at low price, convenience and practicability, global communication, simplified structure, and low energy consumption, instead of blindly pursuing the advancement of technology.

- **Selecting frequency band, address code, and rate.** Bluetooth frequency band is the universal 2.4 GHz industrial scientific medical (ISM), which can be used without applying for permission, thus eliminating the barrier of Bluetooth technology.
  According to IEEE 802 standard, any Bluetooth device can get a public address code, which is a unique 48-bit Bluetooth address code (BD-ADDR). It can be checked both manually and automatically. On the basis of Bluetooth address code, a powerful algorithm with guaranteed performance must be adopted to ensure security. Only in this way can the device identification code obtained be unique. Bluetooth has a data rate of 1 Mbps, performing full-duplex communication in time division mode, and its baseband protocol is a combination of circuit switching and packet switching.

- **Frequency hopping spread spectrum and error correction scheme.** A variety of mobile devices running in the ISM frequency band create huge interference as ISM band is fully opened. The interference is difficult to predict. The working frequency band of devices, such as laptop, Bluetooth speakers, Bluetooth headset, etc., is likely to be in the ISM band. So the transmission error rate in the Bluetooth system is relatively high.
  So far, there are three specific error correction schemes in Bluetooth, namely, 1/3 forward error correction (FEC) code, 2/3 FEC code, and automatic replay request (ARQ). In the Bluetooth system, in order to ensure a fast and effective communication, the number of data retransmissions must be reduced, and the effective way is to use FEC. However, if there is no error code in the sending process, FEC will produce a useless check bit, which will take up a large amount of data bandwidth, resulting in the decrease of data throughput and system efficiency. Therefore, FEC should be carefully chosen under different situations.
  When the packet header with important connection and error correction function is included in the information, FEC must be used in transmission. Specifically, 1/3 FEC method should be adopted for protective transmission to ensure the integrity and reliability of transmission. Otherwise, FEC may not be selected. But in the case of automatic resend, even if there is no serial check bit, if data is sent in one slot, a receipt acknowledgement must be obtained in the next slot. If there is no acknowledgement, an automatic resend request ARQ will be performed.
- **Link types.** In Bluetooth systems, there are mainly two physical links, namely asynchronous connection-less (ACL) link and synchronous connection oriented (SCO) link, differing in the characteristics, performance, and transceiver principle. When transferring some file data or control code, for example, the demands on time are not so urgent while the integrity and reliability of data must be ensured, ACL links are generally used to complete synchronous or asynchronous data transmission. When the demands on time is high, such as using Bluetooth speakers, Bluetooth headset to transmit voice signal, SCO links are adopted to transmit some point-to-point transmission and avoid stuttering.
- **Security mechanism.** Security is the most critical among any communication technology. There are three security modes in Bluetooth technology.

1. **Security mode 1 (not safe).** When the Bluetooth device is in security mode 1, it will not be able to initiate any security program, that is, to send LMP-AuRand, LMP-in-Rand, or LMP-Encryption-Mode-Req.
2. **Security mode 2 (security implemented by the business layer).** A Bluetooth device in security mode 2 does not initiate any security program without receiving L2CAP_ConnectReq – which is a channel establishment request – or sending a channel establishment program. The transmission of security program depends on the security requirements of the requested channel or service. Bluetooth devices in security mode 2 can classify their business security requirements with at least the following attributes: authentication requirements, authentication requirements, and encryption requirements.
3. **Security mode 3 (security implemented by the link layer).** A Bluetooth device in security mode 3 must initiate the security program before sending lMP-Link-Setup-Complete. It can reject connection requests from other hosts based on host settings.

The Bluetooth protocol also classifies devices into trusted ones and untrusted ones according to whether they have been connected in the system. Trusted devices have no limitations at the business level while untrusted ones have.

In the Bluetooth system, the information is encrypted mainly by stream password. The stream password is characterized by good hardware compatibility and the software at the high level of the protocol can manage the key well. Authentication and certification between two devices is a very important part of the Bluetooth system because Bluetooth devices have to be connected, and the most basic mechanism is challenge-response mechanism. Figure 2.15 depicts the Bluetooth legacy authentication.

2.5.2.1 Status of Bluetooth Security Mechanism

Some of the security mechanisms have been developed at the beginning of Bluetooth technology standards, which are designed to ensure the confidentiality, integrity, accessibility, and availability of information. Bluetooth security mechanism provides many application services, such as device authentication, encryption, certification, etc. The link key shared between devices determines the authentication process. During matching, the link key

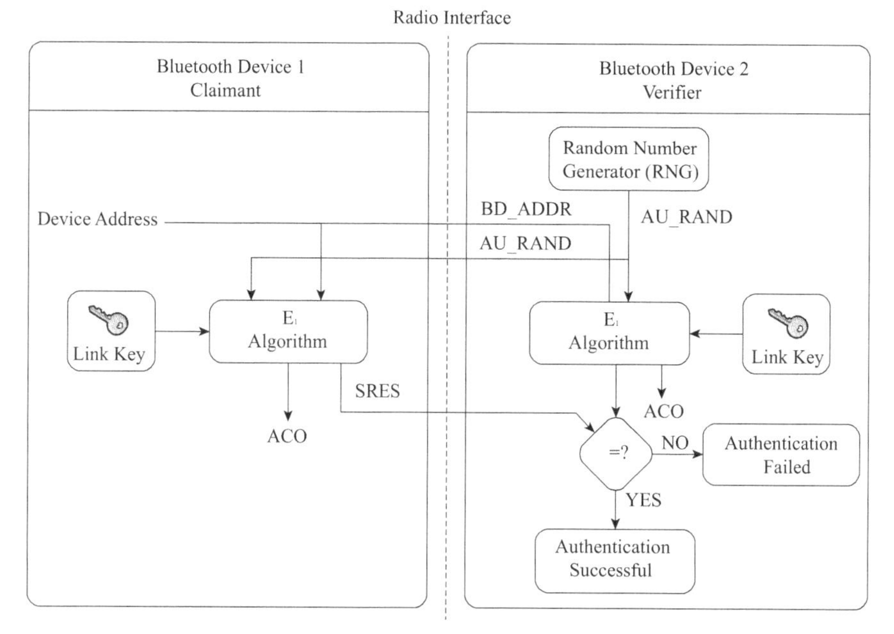

Figure 2.15 The Bluetooth legacy authentication.

can be established by taking the identical PIN code as the initial shared key and loading it into the Bluetooth device. The encryption process of Bluetooth is very simple, mainly to protect the confidentiality of links, and the stream processing method is the commonly used one at present. On the other hand, the ability of two or more devices to connect and transmit is an important function of Bluetooth device authentication. In fact, many defects appeared in the design of Bluetooth security mechanism initially; so it needs to be further improved to meet the requirements of applications with high security.

In Bluetooth security mechanism, there are four different entities at the link layer to maintain security. A 48-bit public Bluetooth address, BD_ADDR, is unique to each device; two user private keys: 128 bits for authentication, 8 bits for encryption; and a 128-bit random number used to distinguish each new processing. The authentication key is used to generate the encryption key, usually when the encryption is activated or when an encryption key is requested. Therefore, it is relatively static compared to the encryption key. Also called the link key, the authentication key is semi-permanent or temporary. Semi-permanent keys are stored in non-volatile storage and can be used after the current session, while temporary keys are only available during the current session.

Another important issue with Bluetooth security is PIN code, which can be either a fixed number or a number chosen by user. The PIN code is between 1 and 16 bytes long and must be entered into both devices during matching. The challenge-response scheme used by Bluetooth devices describes link encryption algorithm using stream cipher, in which LFSR is used. The length of the valid key can be chosen between 8 and 128 bits to accommodate the security requirements of different countries. There are many security flaws of Bluetooth system, and there are also many ways to attack Bluetooth. Table 2.2 lists the existing threats and vulnerabilities of various versions of Bluetooth.

Table 2.2 Threats and Vulnerabilities of Various Versions of Bluetooth.

Classification	Security Threats or Vulnerabilities	Notes
Prior to Bluetooth V1.2	The connection keys based on the unit key are static and are reused in each pair.	A device that uses a unit key will use the same connection key for each device paired with it, which is a serious encryption key management vulnerability.
–	The use of a connection key based on the unit key can lead to eavesdropping and electronic spoofing.	Once the unit key of the device is disclosed (that is, at the time of the first pairing), any other device that has the key can deceive the device or any other device that has been paired with it.
Prior to Bluetooth V2.1	Devices in security mode 1 never activate security mechanisms.	Devices that use Safe Mode 1 are inherently unsafe. For V2.0 and earlier devices, security mode 3 is strongly recommended.
–	The PIN code is too short.	Weak PIN codes (which are used to protect the generation of the connection key during pairing) can be easily guessed, and short PIN codes are prone to be chosen.
–	Lack of management of PIN codes.	It is difficult to set up enough PIN codes in a system with many devices. The best method is to use the random number generator to generate the PIN code for a matching device.

Continued

Table 2.2 *Continued*

Classification	Security Threats or Vulnerabilities	Notes
–	After 23.3 hours of use, the encryption key stream repeats.	The encryption key stream depends on the link key, EN_RAND, the master device BD_ADDR, and the clock, and only the master device clock will change in a particular encrypted connection. If the connection lasts longer than 23.3 hours, the clock value will begin to repeat, producing the same key stream as in the previous connection, which will allow the attacker to learn the original plaintext.
Bluetooth V2.1 and V3.0	The immediate work association model does not provide man-in-the-middle protection during pairing.	The device should require a man-in-the-middle mechanism to protect and refuse to accept unauthenticated connection keys generated by immediate work pairing during the secure simple pairing (SSP).
–	The SSP ECDH key pair can be static.	Weak ECDH key pairs degrade the eavesdropping protection performance of the SSP, allowing the attacker to determine the connection key, and the device should have ECDH key pairs that change periodically.
–	Static SSP keys facilitate man-in-the-middle attacks.	The key can be intercepted by the middleman during pairing.
–	When a device in security mode 4 (that is, V2.1 or later) is connected to a device that does not support security mode 4 (that is, V2.0 or earlier), it is allowed to fall back to any other security mode.	The worst situation is when the device is retreated to security mode 1, which does not provide security.
Prior to Bluetooth V4.0	Attempts to authenticate are repeatable.	Bluetooth devices need to include a mechanism to block an unlimited number of authentication requests.

Continued

Table 2.2 *Continued*

Classification	Security Threats or Vulnerabilities	Notes
–	The master device key for broadcast encryption is shared among all piconet devices.	Sharing keys between more than two parties facilitates a disguised attack.
–	The E0 stream cipher algorithm used in Bluetooth BR/EDR encryption is relatively weak in security.	Federal information processing standard (FIPS) authentication encryption is realized by superimposing application layer FIPS authentication encryption on Bluetooth BR/EDR encryption.
–	Privacy can be compromised if the Bluetooth device address (BD_ADDR) is captured and associated with a particular user.	Once BD_ADDR is associated with a particular user, the user's activity and location may be tracked.
–	Device authentication is a simple challenge/response process for shared keys.	One-way challenge/response authentication is subject to man-in-the-middle attack.
Bluetooth V4.0	low-energy (LE) pair offers no eavesdropping protection.	If successful, the eavesdropper can capture the key allocated during the pairing (i.e., long-term key (LTK), connection signature resolving key (CSRK), identity resolving key (IRK)).
–	Level 1 of LE Safe Mode 1 does not require any security mechanism (i.e., no authentication or encryption).	Similar to BR/EDR security mode 1, this is inherently insecure.
All versions	The connection key may be improperly stored.	If access control is not used to store and protect securely, the connection key may be read or modified by an attacker.
–	The strength of the pseudo-random number generator is unknown.	A pseudo-random number generator may generate static or periodic numbers, which reduces the effectiveness of the security mechanism.

Continued

Table 2.2 *Continued*

Classification	Security Threats or Vulnerabilities	Notes
–	The encryption key length is negotiable.	NIST strongly recommends the use of full 128-bit key strength in BR/EDR (E0) and LE (AES-CCM).
–	There is no user authentication.	The specification provides equipment certification only.
–	There is no end-to-end security mechanism implemented.	Only individual connections are encrypted and authenticated.

2.6 FM Security

2.6.1 Overview of Radio

Vehicle radio is a small radio receiver used mainly for receiving radio programmes. Because of the development of broadcasting, there are many radio waves of different frequencies in the air. In order to select the desired programme and avoid interference, a selective circuit behind the receiving antenna is used to select the desired signal and filter out the noise. The output of the selective circuit is to select the high-frequency amplitude modulation (AM) signal of a certain station. It must be restored to the original audio signal, which is called demodulation, and then sent to the headset. The vehicle radio system is shown in Figure 2.16.

Radios are generally divided into two types, AM and frequency-modulated (FM) radios. The AM radio is composed of input circuit, local oscillator circuit, mixing circuit, detection circuit, gain control circuit, and audio amplifier circuit, etc. The local oscillator signal is mixed with the input signal through an internal mixer. The mixed frequency signal is obtained through the intermediate frequency (IF) selection loop composed of the mid-cycle and the 455 kHz ceramic filter. At this point, the signal of the radio station is converted into an AM wave with an IF of 465 kHz as the carrier. Figure 2.17 shows the working principle of an AM radio.

The input circuit selects a signal from those of a number of radio stations in the air and sends it to the mixing circuit, which converts the high-frequency AM signals from the input into the IF AM ones. The information they carry

Figure 2.16 The vehicle radio system.

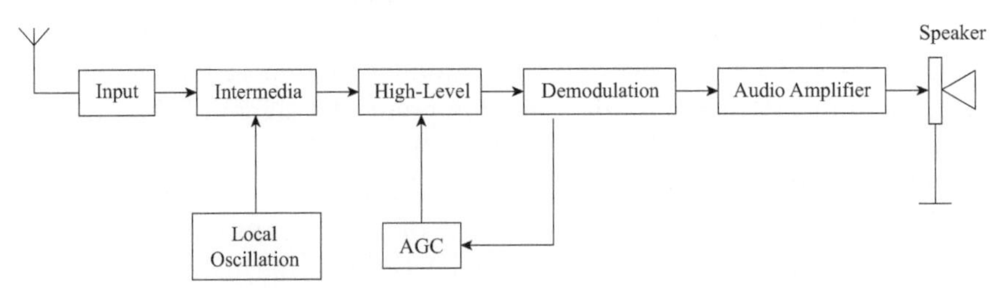

Figure 2.17 The working principle of an AM radio.

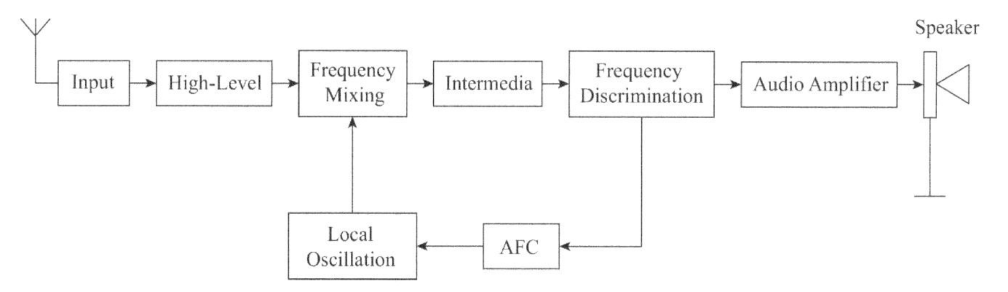

Figure 2.18 The working principle of an FM radio.

remains the same, that is, changing the frequency from high to medium without changing the amplitude.

No matter what frequency the input signal is, it is fixed after mixing, which is stipulated as 465 kHz in China. IF amplifier amplifies the IF signal to the size required by the detector, and the audio signal is detected and transmitted to the audio amplifier. The pre-low amplifier in audio amplifier amplifies the voltage of audio signal and then amplifies it to the level that can push the speaker, and the speaker converts the electrical signal into sound signal eventually.

FM radio is composed of input circuit, high amplifier circuit, local oscillator circuit, mixing circuit, intermediate amplifier circuit, discriminator circuit, and audio amplifier circuit. FM mixing is carried out between the input signal and the local oscillator signal and outputs the FM mixing signal. It is selected by FM IF loop, which is composed of 10.7 MHz filter and IF wave with 10.7 MHz as carrier is extracted. IF modulated wave is amplified by IF amplifier circuit, and then frequency recognition is performed to obtain audio signal, which is output by audio amplification, coupled to loudspeaker, and restored to sound. The process is similar to that of AM radio, as shown in Figure 2.18.

2.6.2 Attack Technique Analysis of FM

Many smart phones and vehicle radios are equipped with embedded FM radio receiver chips. Although the main purpose of embedding these chips is to listen to FM radio stations, the data channel can be utilized by the attacker. Unlike other existing IP-based data channels in portable devices, the data on these radios is open, broadcast, and so far ignored by security providers. In [5], the researchers used FM radio data system protocol as an attack medium to deploy malware and take full control of the victim's equipment.

2.6.2.1 Introduction of FM Radio

The FM standards studied included the radio data system protocol, the U.S. version of which is the Radio Data Broadcasting Standard (RDS). These protocols are widely used by radio stations to transmit data to receivers, including audio programme names, alternative frequencies, and traffic jam updates. RDS carries information about a 57-kHz signal, technically known as a sub-carrier. Each data set is 104 bits long, and each group is divided into four data blocks. According to the number of bits in the group, there are several different types. The group 2A/2B type is called RadioText (RT) and carries arbitrary ASCII text. Type 0A/0B is referred to as the project service (PS) type, and the PS name refers to the current station name. All groups will contain a planned identification (PI) value to the country where the transmission took place. RDS has a data rate of 1187.5 bps, and it takes about 87.6 ms to transfer a group. Also, RDS has a built-in error correction mechanism to detect unit and 2-bit errors in the block.

2.6.2.2 Attack Process of FM Radio

Researchers in [5] took advantage of the RT transferring vulnerability to exploit code or attack scripts. RT can carry arbitrary text data, and researchers can gain control over the target by claiming rights. The attack is carried out in three stages.

- **Stage 1.** As shown in Figure 2.19, the attacker has two main tasks corresponding to different attack chains. First, the attacker creates a malicious application (step 1a) and uploads it to the application market. Then, he gains access to the target device (step 1b). Due to the very low bandwidth of the RDS protocol, packing and compressing the exploit code and attaching the sequence number are needed (step 1c). This grouping step basically splits a binary payload of thousands of bytes into several smaller Base64 encoded packets that are modulated to FM and broadcast via the FM transmitter (step 1d). It does not matter whether the track is transmitted or not because the receiver can receive the RT without playing the audio. If the user is listening to the radio, the attacker can easily transmit the same message as the audio played on a real FM channel.
  There are several ways to determine the transmission frequency. The most effective one is to scan the spectrum of an unused frequency, and another method is to hijack different frequencies of existing channels. Frequency selection depends on the FM capture effect, in which only

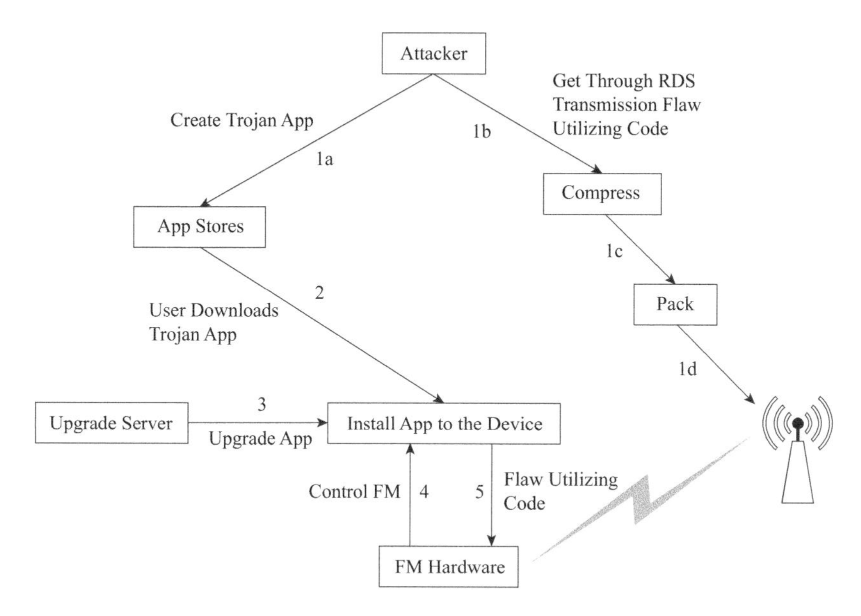

Figure 2.19 An illustration of FM attack process.

one of the two signals which is stronger or closer to the same frequency will be demodulated. The third way is to choose the radio channel to invade. This is a common way to hijack sessions, and its implementation principle is to place the high-power signal transmitter near the target device, but the powerful signal transmitter is located behind the original transmitter, jamming the original signal and sending a whole new set of signals.

- **Stage 2.** The user downloads and installs the Trojan application, making the device a target, and then the application updates the functionality of the device, namely the vehicle radio, by updating the server. The basic function for FM player is playing audio, and updates can add other features. But the attacker can use it to push FM hardware control code into the application, and then the code is decoded in the radio equipment and assembly payload. Then the attacker can load the data with the added DexClassLoader application programming interface (API) class, which allows code to be inserted into a previously installed Android application while running and has the advantage of bypassing any static checks.
- **Stage 3.** Finally, assembling the attack payload, which is executed by the Trojan program to lift weights to control the vehicle radio or other

vehicle systems. Assuming that the attacker has a device capable of broadcasting custom RDS data, in order to send the custom data to the receiver, an FM signal transmitter is required, which can transmit on the authorized frequency. The attacker does not need to use an IP-based communication channel to communicate with the device, and the attacked device does not need to enable the network. A portable signal transmitter is adopted, which is supplied by the 11-V battery. The signal coverage radius of it is up to 3.5 miles, which is large enough to attack a large number of radio devices. The attacker can attack legitimate broadcasts from a radio station and broadcast the custom RDS data to all devices listening to that radio station.

Another assumption is that the attacker releases the application in the application market. In addition to performing its normal functions, this application will also perform the function of the vulnerability program transmitted through RDS, and the API used is typically used by many other legitimate applications to download updates and extension packages when running after installation. Therefore, automated code review mechanisms or behavioural analysis will not be able to detect malicious application tag, and it is also difficult for users to distinguish between real applications and repackaged malicious ones. In order to increase the application downloads, the attacker can choose to repack a well-known application and release in the market.

2.7 GPS Security

2.7.1 Overview of GPS

The GPS is one of the global navigation satellite systems (GNSS) that provides geolocation and time information to a GPS receiver anywhere on or near the Earth where there is an unobstructed line of sight to four or more GPS satellites. Obstacles such as mountains and buildings block the relatively weak GPS signals.

The GPS concept is based on time and the known position of GPS specialized satellites. The satellites carry very stable atomic clocks that are synchronized with one another and with the ground clocks. Any drift from time maintained on the ground is corrected daily. In the same manner, the satellite locations are known with great precision. GPS receivers have clocks as well, but they are less stable and less precise.

Each GPS satellite continuously transmits a radio signal containing the current time and data about its position. Since the speed of radio waves is constant and independent of the satellite speed, the time delay between when the satellite transmits a signal and the receiver receives it is proportional to the distance from the satellite to the receiver. A GPS receiver monitors multiple satellites and solves equations to determine the precise position of the receiver and its deviation from true time. At a minimum, four satellites must be in view of the receiver for it to compute four unknown quantities (three position coordinates and clock deviation from satellite time). Figure 2.20 shows the vehicle-mounted GPS navigation system.

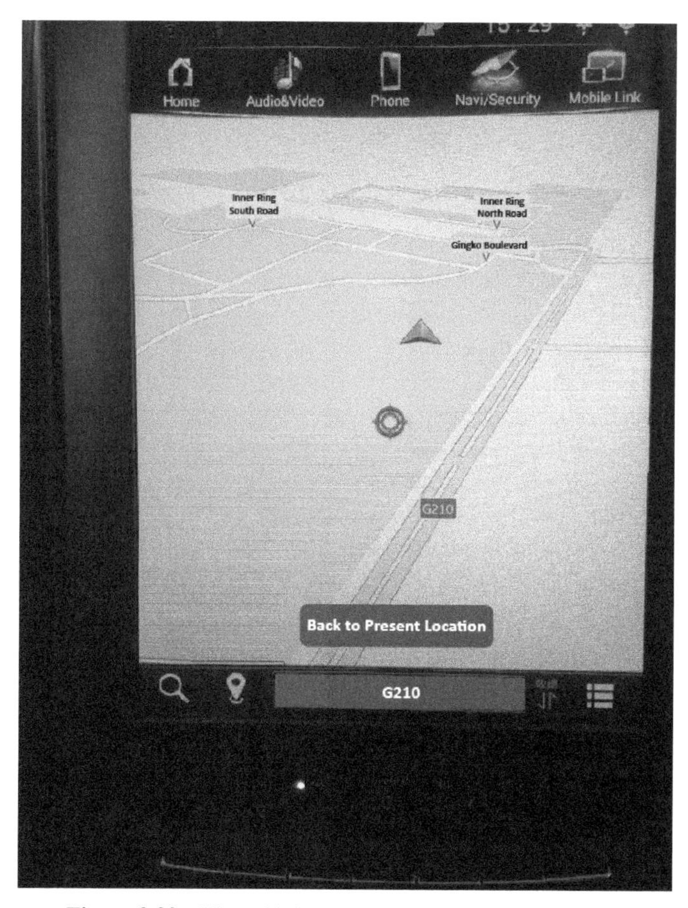

Figure 2.20 The vehicle-mounted GPS navigation system.

The current GPS consists of three major segments, which are space segment, control segment, and user segment.

- **Space segment.**
  The GPS satellite constellation is originally designed to consist of 24 satellites, of which 21 are working ones and 3 are backup ones. The 24 satellites are uniformly distributed in six orbital planes. The planes have approximately 55° inclination (tilt relative to the Earth's equator) and are separated by 60° right ascension of the ascending node (angle along the equator from a reference point to the orbit's intersection). A satellite in one orbital plane is 30° ahead of the corresponding satellite in the adjacent orbital plane to the west. Orbiting at an altitude of approximately 20,200 km and orbital radius of approximately 26,600 km, each satellite makes two complete orbits each sidereal day, repeating the same ground track each day. In this way, even with only four satellites, correct alignment means all four are visible from one spot for a few hours each day.
- **Control segment.**
  It is composed of a master control station (MCS), an alternative MCS, four dedicated ground antennas, and six dedicated monitor stations. For navigation and positioning, GPS satellites are known dynamic points. The satellite's position is based on the ephemeris launched by the satellite – the parameters that describe the satellite's movement and orbit and the ephemeris transmitted by each GPS satellite are provided by the ground monitoring system. Whether the various equipment on the satellite work properly and whether the satellites have been operating in a predetermined orbit, it must be monitored and controlled by ground equipment. Another important role of ground monitoring systems is to keep each satellite in the same time standard, namely the GPS time system. This requires the ground station to monitor the time of each satellite, find out the clock difference, and then send the satellite by the ground injection station, and the satellite is then sent to the user equipment by the navigation message. The ground monitoring system of GPS working satellites includes at least one main control station, three injection stations, and five monitoring stations.
- **User segment.**
  The GPS receiver's mission is to capture satellite signal at a certain satellite altitude, track the operation of the satellite, receive the GPS signal amplification and processing, measure the GPS signal propagation time

from the satellite to the receiver antenna, as well as decipher the GPS satellite navigation message.

The navigation and positioning signal sent by GPS satellite is a kind of information resource shared by countless users. For the vast majority of users on land, sea, and space, as long as the user has a receiving device capable of receiving, tracking, transforming, and measuring GPS signals, namely GPS signal receiver, GPS signals can be used for navigation and positioning measurement at any time. According to the use of different purposes, the user required GPS signal receiver is also different. At present, there are dozens of factories in the world producing GPS receivers and hundreds of products.

In static positioning, the GPS receiver is fixed during the process of capturing and tracking GPS satellites. The receiver measures the propagation time of GPS signals with high precision and uses the known position of GPS satellite in orbit to work out the three-dimensional coordinates of the position of the receiver antenna. Dynamic positioning uses a GPS receiver to determine the trajectory of a moving object. The moving objects on which the GPS signal receiver is located are called carriers (such as ships in flight, airplanes in the air, vehicles walking, etc.). The GPS receiver antenna on the carrier moves relative to the earth in the process of tracking the GPS satellite, and the receiver uses GPS signals to measure the status parameters (instantaneous three-dimensional position and velocity) of the moving carrier in real time.

Receiver hardware and in-machine software as well as GPS data post-processing software package constitute a complete GPS user equipment. The structure of GPS receiver is divided into two parts: antenna unit and receiving unit. For geodesic receiver, the two units are generally divided into two independent components. During observation, the antenna unit is placed on the measurement station, and the receiving unit is placed in an appropriate place near the measurement station. The two units are connected into a complete machine by cable. Others make the antenna unit and the receiving unit into a whole and place them on the test site during observation.

2.7.2 Attack Technique Analysis of GPS

GPS signal spoofing consists of broadcasting fake signals over the real GPS signals in order to take control of a GPS receiver that will continue to track

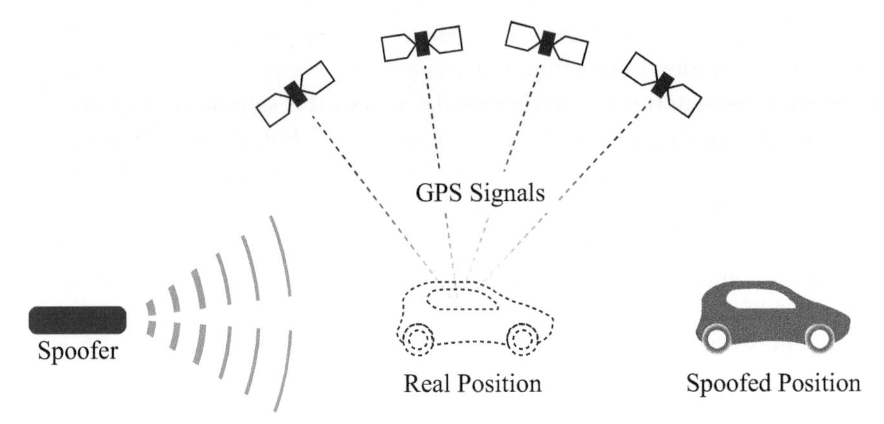

Figure 2.21 Spoofing a vehicle to introduce an erroneous result for position.

those signals in error. GPS is very sensitive to this type of attack due to the weakness of satellite signals at the earth's surface and the fact that these signals are public and not protected as such. A successful spoofing attack consists of taking control of a receiver without it noticing the attack. The goal of the attack is to introduce an erroneous result for position or timing or both.

Spoofing differs from jamming in that a successful attack is not noticed by the receiver. A successfully jammed receiver will typically lose its calculated position/speed/time (PVT) result, which is easily detected by the equipment itself or the master system.

With critical systems, solutions exist for addressing a missing GPS result. Jamming is therefore relatively well accounted for in practice, without critical impact. However, a successful spoofing attack can induce errors in the receiver without any protective mechanism taking notice. It is easy to imagine the potentially disastrous consequences of spoofing on an airliner, a banking system, or an electrical grid.

As a first step, a spoofing resistant receiver should therefore detect the attack and ideally, in the second step, be able to recover the genuine GPS calculation result. Unfortunately, the protection of GPS receivers was treated as an afterthought for a long time because of the increased complexity involving spoofing attacks when compared to jamming. Since the turn of the century, an effective, evolved spoofing attack has required the procurement of GPS constellation simulators costing hundreds of thousands of dollars or developing complex equipment yourself. Today, the growing popularity of cheap and powerful software defined radio (SDR) equipment makes this a clear and present danger.

Every manufacturer or integrator of GPS receivers should therefore include robust spoofing resistance testing as part of its certification process. These tests should be performed with the aid of a GPS simulator because it is not feasible to broadcast false GPS signals in the open sky. Thanks to its ability to synchronize multiple radios and sessions, Orolia's Skydel GPS simulator is designed to address these types of scenarios.

This application note describes the procedure for performing a spoofing attack, in a controlled environment, on a vehicle in order to test the receiver's robustness in such a scenario. Other tests could have been performed, for example, testing the robustness of a time-based spoofing attack.

2.8 Experiments

2.8.1 Attack Experiment Against RKE

The RKE system is composed of a controller, a RF transmitter in the remote control key, and a receiver end at the car. The controller at the receiver end can be used as the control module of the receiver, and some other functions can be integrated. According to the complexity of the controller functions, 8-, 16-, or 32-bit microcontrollers can be selected.

Nowadays, in most of the vehicles, the RKE system uses one-way protocol to achieve cyclic code verification, and there is no receiver at the key fob; so it has the risk of being attacked. This section will introduce an RKE attack experiment, which uses the vulnerability of one-way protocol in RKE system to illegally open the door lock through replay attack.

2.8.1.1 Experimental Environment

HackRF One is adopted from Great Scott Gadgets in this attack experiment, which is a SDR peripheral, capable of transmission or reception of radio signals from 1 MHz to 6 GHz. It covers many licensed and unlicensed ham radio bands. HackRF One can be used to sniff out radio waves emitted from keys and replay them to vehicles for a simple RKE attack.

As for software environment, GNU radio is a free and open-source software development toolkit that provides signal processing blocks to implement software radios. It can be used with readily available low-cost external RF hardware to create SDRs, or without hardware in a simulation-like environment. In the experiment, GNU radio is used to record and simulate the received radio waves, which is an important part of replay attack. This attack experiment against RKE system is depicted in Figure 2.22.

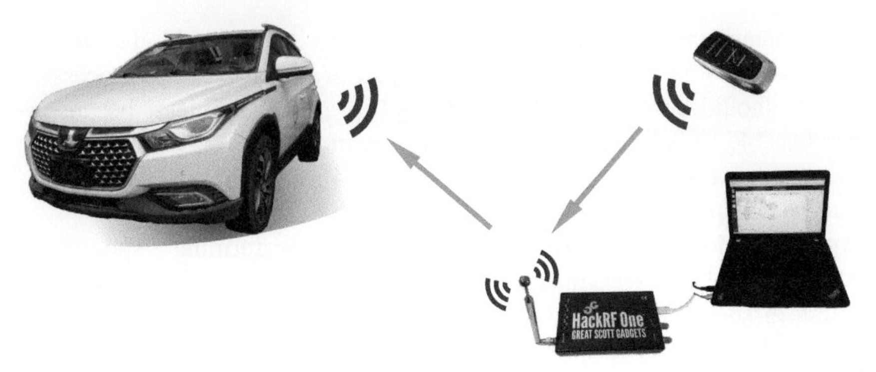

Figure 2.22 A demonstration of the attack experiment against RKE system using hackRF One.

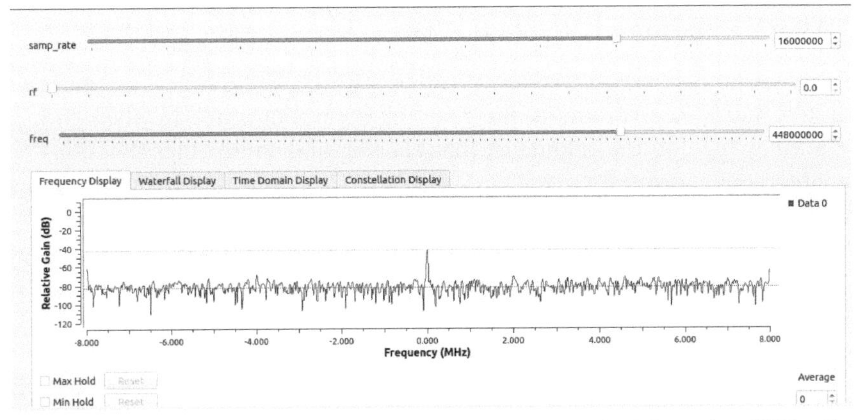

Figure 2.23 The radio signal from a key fob recorded by hackRF One using GNU radio.

2.8.1.2 Experimental Method

Before the experiment, it is necessary to keep the car key away from the vehicle and try to find a quiet environment to collect the radio signal from the key. Figure 2.23 shows the radio signal from a car key recorded by hackRF One using GNU radio.

Since the radio signal is transmitted in the way of cyclic code, the signal sent each time will be different. Therefore, in order to better realize the RKE attack, the attacker can build a radio signal library to store a large number of radio signals sent by the key. After the signal collection, the attacker can use hackRF One to replay the car. The radio signal of replay attack is shown in Figure 2.24.

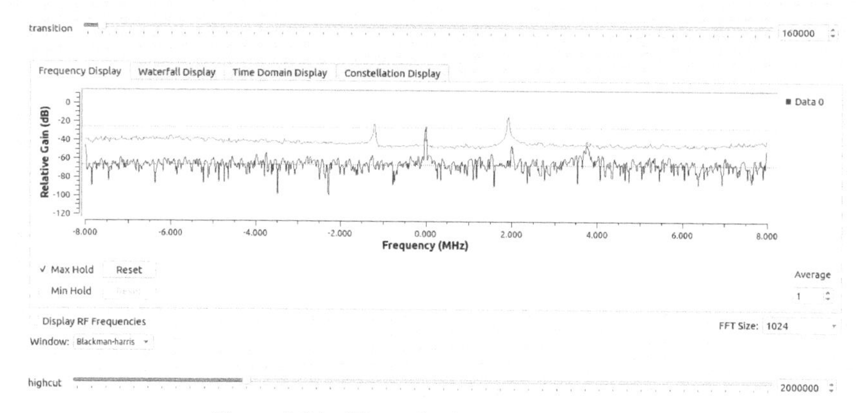

Figure 2.24 The radio signal of replay attack.

The RKE has brought a good user experience to people and met the requirements of convenience and comfort. However, due to the technical limitations of RF one-way communication, it has its own shortcomings in security. For this reason, in order to better bring users safe and convenient services, automotive manufacturers began to use RF two-way communication technology to achieve security verification.

2.8.2 Attack Experiment Against TSP

Nowadays, the remote service function in most of the vehicles is composed of TSP and T-BOX. TSP mainly includes vehicle equipment manufacturers, content providers, network operators, and automotive manufacturers. Equipment manufacturers are responsible for providing TSP hardware and software support for automotives, while content providers offer users with GPS navigation entertainment and multimedia information services. Network operators include mobile operators, fixed line operators, and satellite ones, with the aim of providing communication infrastructure between ICV and TSP. Automotive manufacturers are responsible for real-time monitoring of vehicle anomalies and protecting the safety of users.

In addition, T-BOX, as the core equipment of communication between automotive and TSP, uploads vehicle status information and user information to TSP to realize cloud tracking and protection of vehicles. At the same time, T-BOX can also receive and process control instructions from TSP. TSP provides users with more convenient, safe, and reliable services by remote control of car door switch ignition/flameout and boot opening. However, once

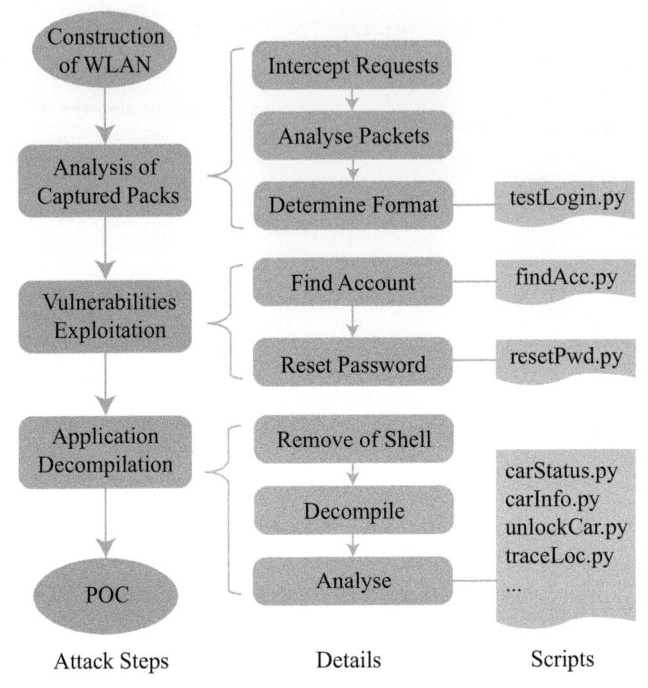

Figure 2.25 The process of TSP attack.

TSP is broken, users' privacy and in-vehicle data will be stolen by malicious attackers, and even the target vehicle will be completely controlled by the attacker.

This section will introduce a TSP attack experiment [6]. The authors found the verification vulnerability of TSP cloud in Luxgen and verified that there is a risk of leakage of TSP cloud user data. Moreover, the attacker successfully logged in to the TSP server without authorization and control the vehicle by sending false instructions to TSP. The process of TSP attack is shown in Figure 2.25.

2.8.2.1 TSP Attack Process

The attack process against TSP can be summarized as follows.

- **WLAN network construction and data packet crawling and analysis.** In order to sniff the data packets between TSP and Luxgen cloud control app, it is necessary to build a WLAN by using wireless router. Then, all request–response packets are sniffed by using Ettercap or

Burpsuite in the WLAN, and some unencrypted data in the packets are analysed. Finally, Python scripts are written to determine the format of the app login request.

- **User name and password cracking.** After understanding the correct request login format of app, the problem of user name and password cracking is to be solved. It is found that when an attacker inputs a registered user name, the cloud will return a response of password error or successful login. When the user name does not exist, the cloud will return a response that the user name does not exist. According to different feedbacks, a large number of mobile phone numbers can be used to request login in order to find the mobile phone number of the registered Luxgen cloud control app. Because the process of the experimental attack is illegal, only the mobile phone number bound to the experimental vehicle is used in the experiment to verify the feasibility of the attack scheme. In real life, if the owner places a mobile phone with a SIM card in the car, the attacker can use the mobile phone to find the user name of the target vehicle.
  Knowing the user name of the target vehicle, the next step is to figure out the correct login password and the frame number bound with the user name so as to successfully login to the car app. In the experiment, sniffer tool was used to obtain the car frame number information, and the password reset vulnerability was used to change the login password. When receiving the password resetting request, TSP cloud will send a six-digit check code to the bound mobile phone. The digital check code is valid within 5 minutes, and the input error will not affect the re-input. Therefore, the software tool can be used to repeatedly send a six-digit check code to the TSP terminal until the verification is successful to enter the password change interface.
- **Acquisition and replay of control instructions.** When Luxgen cloud control app sends commands to the cloud, such as opening or closing the door, ignition or flameout, whistle, etc., the sniffer tool can accurately capture the corresponding request instructions. Then, user login scripts and command control pins are written in Python language. MITM attack is used to send login requests and control requests to the cloud so as to realize the attack on Luxgen automotive.

2.8.2.2 Experimental Results of TSP Attack

Through the experiment described above, it can be found that there is a risk of user privacy leakage between the app and the cloud. The results of TSP

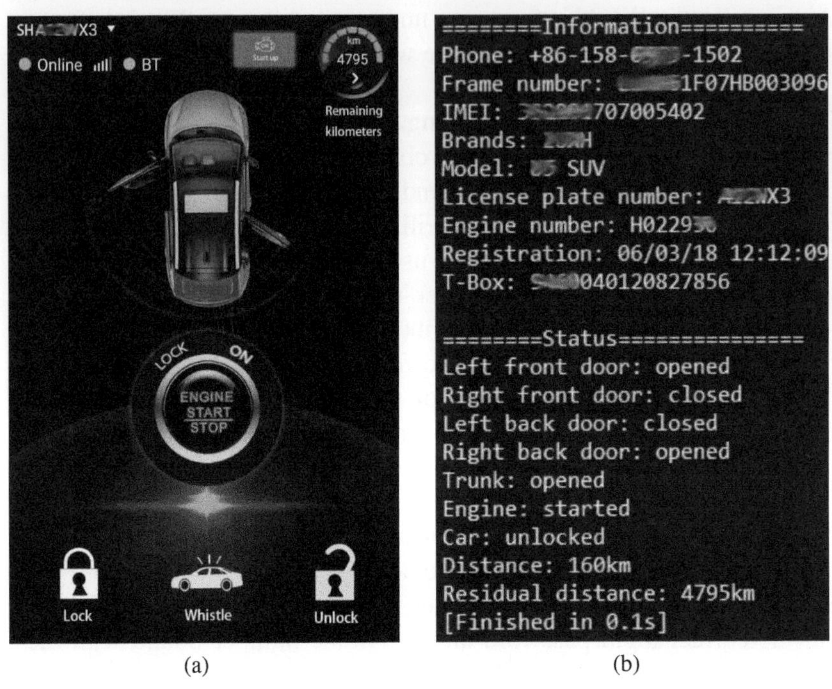

Figure 2.26 The result of infiltrating TSP. Privacy information and current operating status of the vehicle are obtained from the malicious scripts: (a) the status of the car officially shown; (b) privacy exposed by the attack scripts.

attack experiment are shown in Figure 2.26. Figure 2.26(a) shows the status of Luxgen cloud control app, while Figure 2.26(b) shows the illegally obtained vehicle privacy information. Figure 2.27 shows the current vehicle location information illegally obtained. In addition, the attacker can also illegally steal the user name and password of the app and control the opening/closing of vehicle door, ignition/flameout, and whistle.

2.8.3 Attack Experiment Against TPMS

TPMS can alarm when the tire appears high pressure, low pressure, and high temperature, remind the car owner to pay attention to driving safety, and provide safety guarantee for people's travel. Because direct TPMS can more accurately reflect the internal conditions of tires, the NHTSA requires that all new vehicles sold or produced in the United States must be equipped with direct TPMS since 2008. However, due to the defects of the TPMS

Figure 2.27 The position information of the car which is sent to the user's phone through the TSP: (a) the current car's location officially shown; (b) real-time tracking realized by malicious scripts.

proprietary protocol, there is a risk of being attacked. Malicious attackers continue to send low tire pressure signals to the car to stop and monitor tire conditions, which may provide ample opportunities for pranks and criminal activities. This section will introduce a typical attack experiment about TPMS [8], hoping that more researchers will pay attention to the security of ICVs and providing suggestions for the safety protection of vehicles.

2.8.3.1 Structure of TPMS

A typical TPMS consists of the following components: a TPM sensor mounted behind each tire stem, a TPM ECU, a receiving unit (integrated or independent of ECU), an instrument panel warning light, and one or four skylines connected to the receiver unit. The TPM sensor transmits the

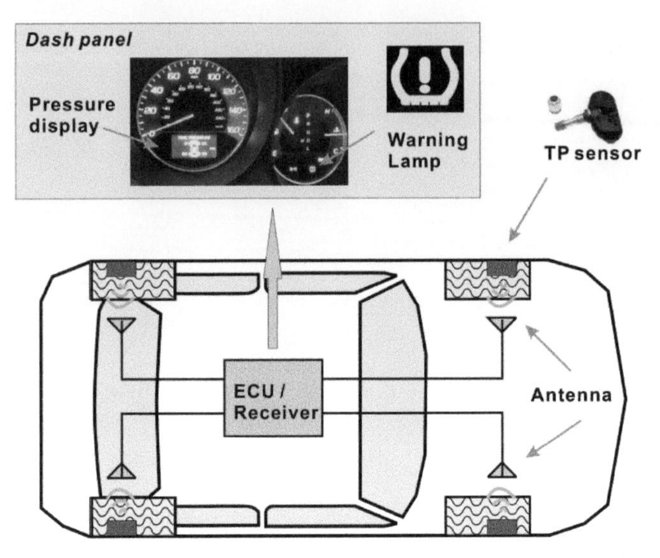

Figure 2.28 A typical TPMS architecture with four antennas [8].

temperature and pressure it detects along with its ID, while the TPM receiver receives the data from the sensor and sends it to the instrument panel to display the TPM. The structure of TPMS is shown in Figure 2.28.

Because of the proprietary communication protocol between TPM transmitter and receiver, the data transmission usually uses 315 or 433 MHz band (UHF) and ASK or FSK modulation. With the characteristics of broadcast transmission and no encryption authentication, TPMS is easy to be used by hackers.

2.8.3.2 Reverse Engineering TPMS Communication Protocols

The key of penetration attack needs to start from the protocol of specific sensors. Therefore, we need to know the modulation scheme, coding scheme, and information format of the protocol used by TPM sensor in detail. In addition, we also need to be familiar with the decoding scheme as well as the method of sniffing sensor information. Through these means, reverse engineering research on TPM sensor is carried out.

The tools include the ATEQ VT55 TPMS trigger tool, two tire pressure sensors (TPS-A and TPS-B), a low noise amplifier (LNA), and one laptop connected with a universal software radio peripheral (USRP), with a TVRX daughterboard attached. These are used to sniff the data from the TPM pressure sensor and reverse its protocol to crack the information such as

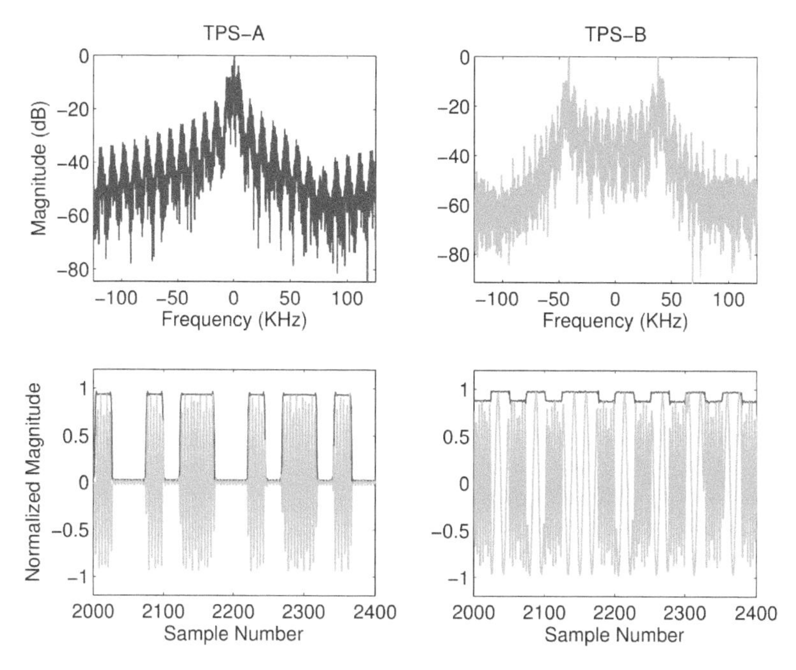

Figure 2.29 The comparison of FFT and signal strength time series between TSP-A and TSP-B sensors [8].

ID number and data contained in the signal. The comparison of fast Fourier transform (FFT) and signal strength time series between TSP-A and TSP-B sensors is shown in Figure 2.29.

Through the reverse engineering of TPMS communication protocol, the message sent by TPM sensor can be obtained successfully. Because the protocol is not encrypted, the attacker can forge the ID and data content according to the message of the cracked sensor, thus successfully realizing the TPMS attack experiment. The experimental results are shown in Figure 2.30. Malicious sending 0 psi information makes the tire pressure display abnormal and the warning light is on. The details of the experiment can be found in [8].

2.8.4 Attack Experiment Against Vehicle Bluetooth via BlueBorne Vulnerability

Armis labs exposed a new attack vector, named "BlueBorne", which endangers major mobile, desktop, and Internet of Things (IoT) operating systems,

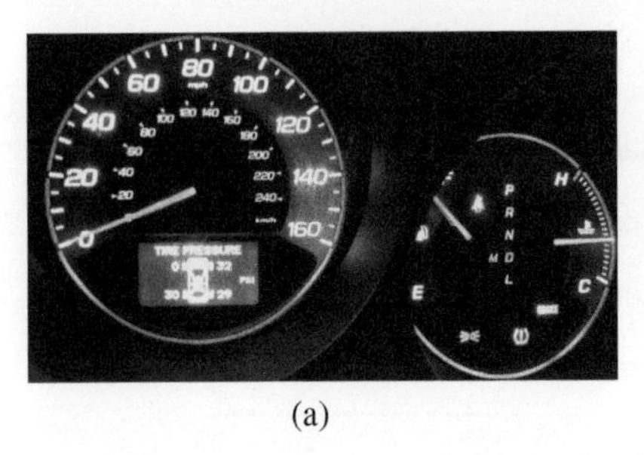

(a)

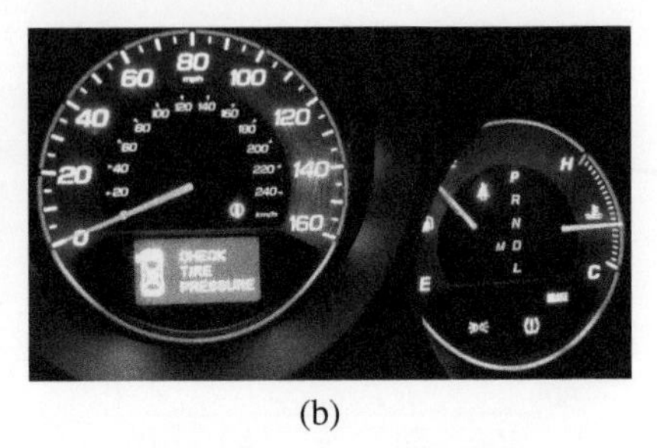

(b)

Figure 2.30 Dash panel snapshots: (a) the tire pressure of left front tire displayed as 0 psi and the low tire pressure warning light was illuminated immediately after sending spoofed alert packets with 0 psi; (b) the car computer turned on the general warning light around 2 seconds after keeping sending spoofed packets [8].

including Android, iOS, Windows, Linux, and the devices using these operating systems [1].

Armis revealed eight related Bluetooth vulnerabilities, four of which were classified as critical. BlueBorne attack can control the attacked device, access the related network and database, implant malware, and spread it horizontally to adjacent devices. Among the attack devices that armis flawed, it included the car audio system. Once the car audio is maliciously used by hackers and implanted with virus, the attacker can not only steal the driver's private information but also send control instructions to CAN bus via the car audio system, which seriously threatens the safety of drivers and passengers. The eight Bluetooth vulnerabilities are as follows:

1. Linux kernel remote code execution (RCE) vulnerability – CVE-2017-1000251;
2. Linux Bluetooth stack (BlueZ) information leak vulnerability – CVE-2017-1000250;
3. Android information leak vulnerability – CVE-2017-0785;
4. Android RCE vulnerability #1 – CVE-2017-0781;
5. Android RCE vulnerability #2 – CVE-2017-0782;
6. The Bluetooth Pineapple in Android – Logical Flaw CVE-2017-0783;
7. The Bluetooth Pineapple in Windows – Logical Flaw CVE-2017-8628;
8. Apple Low Energy Audio Protocol RCE vulnerability – CVE-2017-14315.

2.8.4.1 What is BlueBorne?

BlueBorne is an attack vector that hackers can use for Bluetooth connection penetration and control the target device. The attacker does not need to pair the target device with the attacker's device or even set it to discoverable mode. The attacker can combine the detected Bluetooth vulnerabilities to achieve a devastating effect on the target device. BlueBorne vulnerability can be used to attack vehicle audio equipped with Bluetooth connection. It aims at the weakest link in network defence and the only link without security measures. Moreover, it can also be transmitted from one car to another through the air, which is highly infectious. In addition, because Bluetooth processes have high privileges on all operating systems, this is what hackers would like to see. Hackers can use their privileges to attack target test vehicles wantonly, such as data theft, extortion software, vehicle control, and even create large-scale botnets to paralyse the vehicles.

2.8.4.2 Process of BlueBorne

The workflow of using BlueBorne attack vector to invade the vehicle is divided into the following stages. First, the attacker locates an active Bluetooth connection in vehicle around him or her (the device can be recognized even if it is not set to discoverable mode). Next, the attacker obtains the MAC address of the Bluetooth device in the vehicle, which is the unique identifier of the specific device. By detecting the device, the attacker can determine which operating system his victim is using and adjust his utilization accordingly. Then, the attacker can gain the access rights to attack the malicious vehicle by exploiting the vulnerability of Bluetooth protocol in related platforms. In this stage, the attacker can choose to create a MITM attack and control the communication of vehicle equipment, or implant virus into the vehicle device, and send control instructions to the vehicle bus to control the vehicle state.

Bluetooth attacks can spread in the air and between devices, posing a huge threat to any organization or individual. For more detailed technical documentation on BlueBorne, please refer to the technical document mentioned in [1].

2.8.5 Attack Experiment Against Road Navigation System

Nowadays, GPS has become an essential equipment in vehicle travel, which can provide accurate and reliable location information for free. However, while GPS brings convenience to users, there are also some security defects. This section will introduce an attack scheme for the road navigation system. The researchers inject malicious code into the map application to forge the GPS position remotely to induce the attacked vehicle to arrive at the wrong destination or pass the wrong path [3].

2.8.5.1 Goals of Road Navigation System

According to the reverse engineering of the target road navigation system app, a new attack scheme of road navigation system is proposed. Specifically, attackers inject malicious code into the target map application to realize path guidance by manipulating target map app (such as Navidog and Google Maps) in different network environments (such as Wi-Fi, 3G/4G). The attacker injects malicious code into the target map application, wiretaps the victim's privacy information remotely, and changes the longitude and latitude of the driving route in the path planning function, which can achieve three purposes: wrong destination attack, malicious waypoint attack, and real-time tracking attack.

Wrong destination attack: attackers tamper with the final destination and guide the attacked vehicle to the wrong destination. As shown in Figure 2.31(a), the initial destination of the victim is B. If the malicious attacker secretly modifies the longitude and latitude of the destination to point C, the victim is likely to reach point C instead of point B.

Malicious waypoint attack: the attacker can also change the path of the victim without changing the destination. As shown in Figure 2.31(b), the

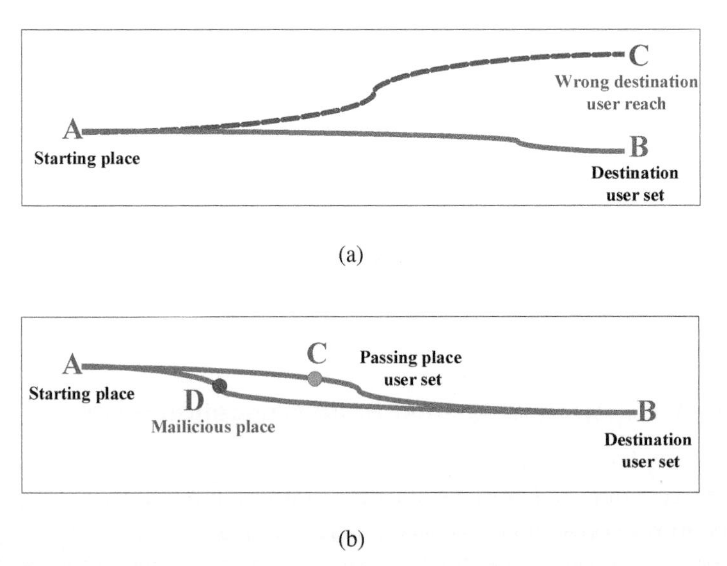

Figure 2.31 Two goals of our threat model. (a) Modify the longitude and latitude of destination before navigation map app plans route to manipulate the victim's destination and route. (b) Change the longitude and latitude of waypoint before navigation map app plans route to manipulate the victim's waypoint and route.

victim passes through point C when he reaches point B, and the attacker can change it to pass through point D without knowing it.

Real-time tracking attack: as long as the victim starts the navigation software, the attacker can use the embedded trojan horse program to track the victim's position in real time.

2.8.5.2 Detailed Attacking Process

After in-depth research, the researchers found that many navigation map applications have loopholes. If hackers use these vulnerabilities to invade the navigation system and implant malicious viruses, once the navigation system is running, the attacker can track the victim's location in real time and make malicious attacks. The detailed steps of navigation map application intrusion are shown in Figure 2.32.

In the research of navigation map app, the shell of the target APK should be removed first. Then, signature verification needs to be bypassed so that the tampered APK can be repackaged. Next, we need to analyse the decompiled code and find out the core code and destination setting code of the map navigation path planning part from the code. After finding the key location code, malicious programs can be implanted to control its core functions. Finally, the APK embedded with malicious attack code is repackaged to develop new attack functions for malicious attack. The specific attack steps can be found in [3].

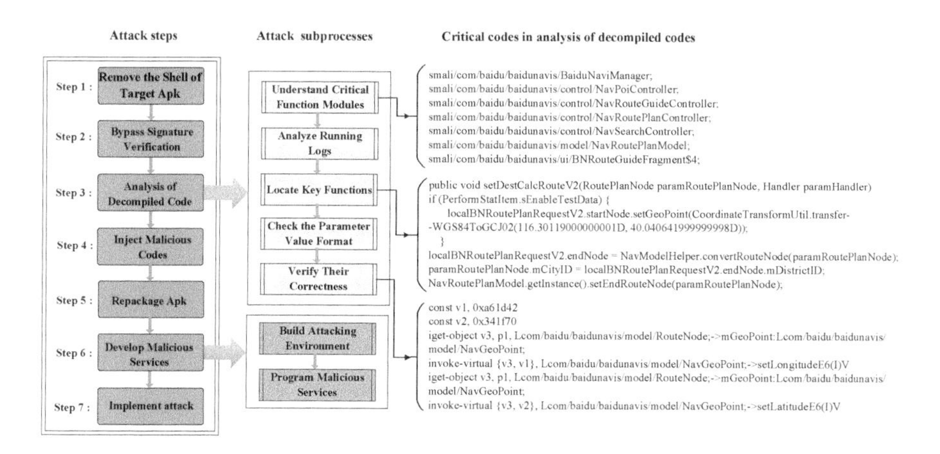

Figure 2.32 The detailed steps of navigation map application intrusion.

2.8.5.3 Experiments and Results

Figure 2.33 shows the experimental results of malicious intrusion by researchers. They forge a malicious server to deceive the navigation system

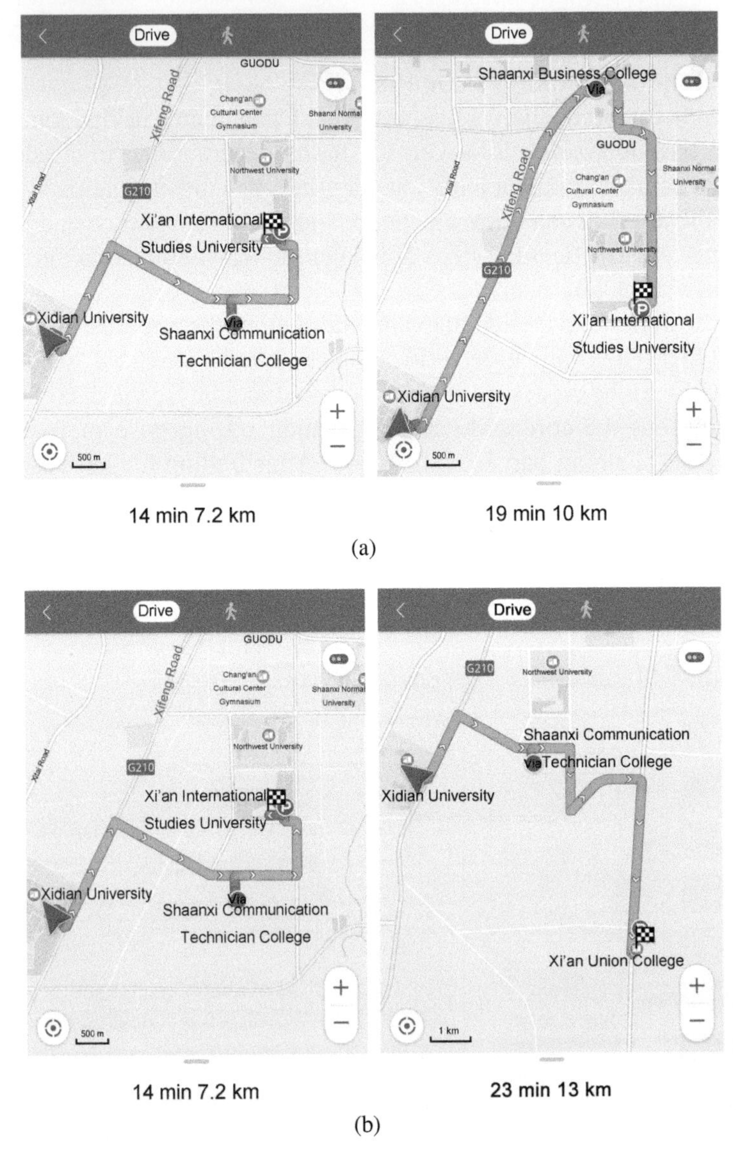

Figure 2.33 The experimental results of Navidog routing falsified by remote spoofing attack.

and make the target application receive false response from the server. Then the malicious attacker can inject the key parameters into the application through the server for malicious attack. Figure 2.33(a) shows the original recommended route (left) and the route tampered with after malicious attack (right). Figure 2.33(b) shows that the researchers maliciously tampered with the destination of the vehicle.

3

Unmanned Driving Security and Navigation Deception

3.1 Basic Introduction to Unmanned Driving

3.1.1 What is an Unmanned Vehicle?

An unmanned vehicle, also known as an autonomous vehicle, driverless car, or robotic car, is a cutting-edge technology for safe and effective driving without human manipulation. It relies on artificial intelligence, computer vision, monitoring devices, as well as positioning and navigation technologies. Figure 3.1 shows a representative self-driving car.

According to the Society of Automotive Engineers' (SAE) definition, unmanned driving can be classified as six levels. At level 0, the vehicle is operated entirely by the driver. Level 1 means that the vehicle can assist the driver in certain driving tasks under certain circumstances. At level 2, the unmanned vehicle can perform certain driving tasks, but the driver needs to monitor the surrounding environment at all times and be ready to take over in case of danger, which is a level that many vehicles have reached. At level 3, the driver is almost ready to take over the vehicle at all times, and the vehicle can perform all actions independently. Level 4 and level 5 are fully unmanned driving, in which vehicles are no longer controlled by drivers. The difference between them is that vehicles at level 4 can be completely independent only under certain conditions such as highways, while vehicles at level 5 can be completely independent under any conditions.

In 1999, the self-driving car Naclab-V developed by Carnegie Mellon University completed its first test. Since then, unmanned vehicles have gradually drawn the world's attention, and many laws and regulations have been introduced for the experiment of unmanned cars on the open road. At present, with Internet of vehicles (IoV), advanced sensor technologies, and artificial

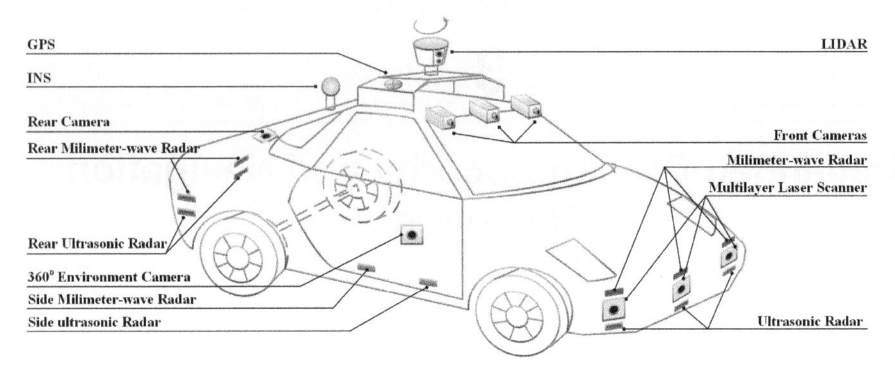

Figure 3.1 Illustration of a representative unmanned vehicle.

intelligence algorithms, unmanned driving is becoming more and more practical and commercial. Leading companies such as Google and Uber have developed unmanned cars that can travel smoothly on the open road. Tesla has achieved mass production of level-3 self-driving cars. Bavarian Motor Work (BMW), Mercedes-Benz, Volkswagen, Volvo, and other traditional automotive companies are also stepping up their efforts to develop unmanned driving technologies.

With the increasing number of automobile users, the problems of traffic congestion and road accidents are becoming more and more serious. The unmanned vehicle can plan and coordinate the travel route, thus greatly improving the travel efficiency and reducing the energy consumption to a certain extent. It can also help avoid drunk driving, fatigue driving, and other safety risks, as well as reduce drivers' mistakes in the process of feeling, judgement, or operation. Fortunately, the attention and investment from all over the world have accumulated a lot of experience for unmanned vehicles, and the unmanned driving technology is showing a trend of rapid development, which is expected to bring more safety, convenience, and comfort to people's life.

3.1.2 Core Functional Modules of Unmanned Driving

To achieve unmanned driving, vehicles need to see, think, and act on their own. Corresponding to these three objectives, unmanned cars need to realize several main functional modules, namely, perception, planning, and control, which are shown in Figure 3.2 and will be introduced in detail in the following.

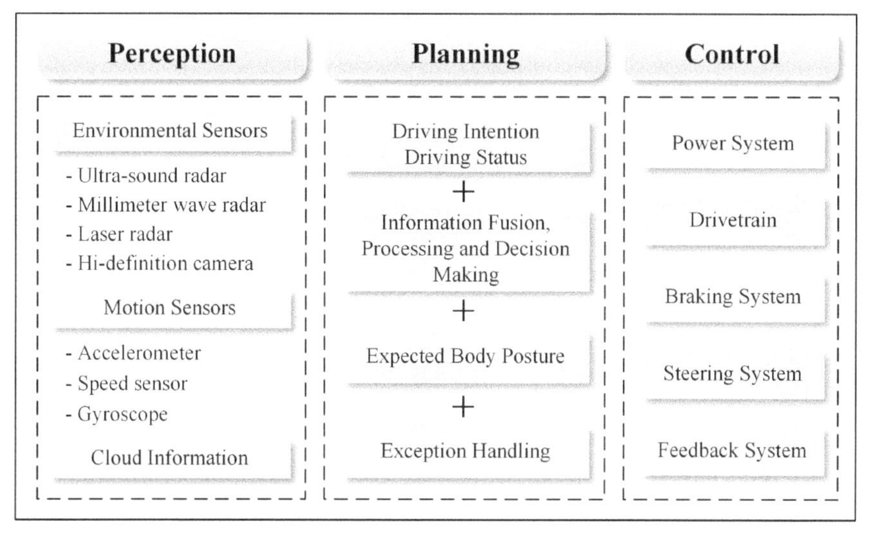

Figure 3.2 Three main functional modules in unmanned driving.

3.1.2.1 Perception

The unmanned car needs to obtain and process the environmental information in real time. Judging from most of the current technical schemes, light detection and ranging (LiDAR) can achieve 60%–75% of the environmental information acquisition and realize three-dimensional spatial perception of the surrounding environment, followed by the image information obtained by the camera, the distance information obtained by millimeter wave radar, the vehicle position and attitude information obtained by global positioning system (GPS) positioning and inertial navigation, and, finally, the various environment information obtained by other photoelectric sensors such as ultrasonic sensors, infrared sensors, and so on. Generally speaking, the more kinds of sensing devices there exist, the more expensive the price is, and the higher accuracy and the larger recognition range it can achieve. However, each sensing device has its limitations.

Radar has strong robustness to jamming factors such as illumination and colour. LiDAR, millimeter wave radar, and ultrasonic radar all have their own advantages. However, no matter how many kinds of radar are installed and how high the sampling rate is selected, it is impossible to completely solve the problem of pit reflection, smoke and dust interference, as well as the impact of bad weather conditions such as rain, snow, and fog. It is also difficult to achieve a real all-weather, all-day, all-space dimension detection. Therefore, radar is not perfect.

No matter which camera is used, such as monocular camera, binocular camera, multi-eye camera, or depth camera, and no matter how clear the pixel is and how high the sampling rate is, the problems of image acquisition and processing cannot all be solved. Due to the diversity and complexity of the road environment, weather condition, and the motion characteristics of the self-driving vehicle itself, the camera is vulnerable to many uncertain factors, such as illumination, angle of view, scale, shadow, defacement, background interference, and target occlusion. Also, in the driving process, the lane line, traffic lights, and other traffic elements can have a certain degree of wastage; so the camera is easy to be disturbed, and there is no completely ideal camera sensors.

The positioning and navigation system provides high precision and reliable positioning, navigation, and timing services for unmanned driving. The carrier phase difference technology and inertial navigation system have laid an important foundation for real-time accurate positioning and position accuracy maintenance. However, no matter how good the location service is, and how high the precision of gyroscope is, the positioning system still has the problems of insufficient sampling frequency, too complex geographical environment, too long initialization time, and failure of satellite signal; so the positioning and navigation system always has defects.

For different driving tasks, different types of sensing devices are needed. Figure 3.3 illustrates the application of various sensors for unmanned driving. It can be seen that all kinds of sensors have their own detection range and application scenarios. In addition, Table 3.1 lists the technical comparisons of

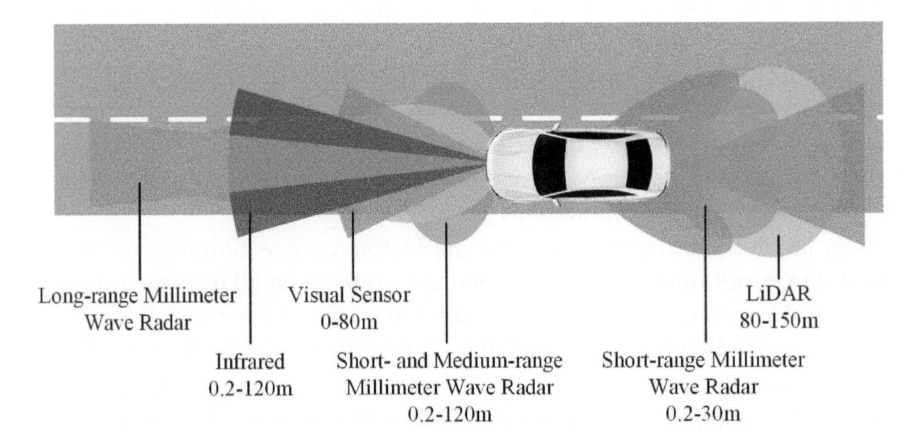

Figure 3.3 Applications of various sensors in unmanned vehicles.

Table 3.1 Technical Comparison of Main Vehicle Sensors.

Type	Ultrasonic Radar	Millimeter Wave Radar	Hi-definition Camera	LiDAR
Detection Range (m)	15	1000	–	300
Velocity Range (km/h)	<100	>1000	–	>300
Radial Motion	Good	Good	–	Good
Tangential Motion	Bad	Bad	–	Bad
Angle Measurement	Good	Good	–	Very good
Influencing Factor	Wind, dust, etc.	Bad weather	Light	Rain, snow, etc.
Cost	Low	Medium	Medium	High
Penetrability	Bad	Good	Bad	Bad

on-board sensors such as ultrasonic radar, millimeter wave radar, hi-definition camera, and LiDAR. The detection range, velocity range, and other technical features are summarized in detail. It is not necessary to configure the most complete and expensive sensing devices to accomplish the driving tasks but to select the appropriate sensing devices and combine them together to realize the optimal configuration.

3.1.2.2 Planning

Planning is one of the core functions of unmanned driving. It first fuses multiple sensing information and makes task decisions according to driving requirements. Then, under the premise of avoiding possible obstacles, through some specific constraints, it plans multiple optional safety paths and selects an optimal path from these paths as the vehicle trajectory. The planning function can be divided into global planning and local planning. Global planning is the process of planning a collision-free optimal path under some specific conditions by obtaining map information, while local planning is based on global planning and some local environmental information to avoid bumping into unknown obstacles and finally to reach the destination.

The development and integration of the planning module can be divided into four key links, that is, information fusion, decision making, trajectory planning, and exception handling. Among them, information fusion completes multi-sensor data association and fusion to establish peripheral environment model; decision making completes the global path planning task of intelligent vehicle; trajectory planning carries out trajectory planning of unmanned vehicles in different local environments; exception handling is responsible for unmanned vehicle's fault warning and safety mechanism. The following is a brief introduction of these four key links.

- **Information fusion.** In environmental perception, a variety of sensors are usually used to collect and analyse the driving environment data, which can be divided into three sources, that is, environmental sensor, positioning and navigation equipment, and vehicle-to-everything (V2X) communication equipment. When unmanned vehicles travel in complex and changeable road conditions, uncertainty about the surrounding environment is likely to put them at risk, especially when they only rely on a single environmental sensor. Therefore, using multiple sensors and data fusion to percept the surrounding environment can fully and accurately describe the characteristics of the target object and reduce ambiguity as well as improve the accuracy and robustness of the planning module. Data fusion technology includes data conversion, data association, fusion calculation, and so on. The core of data fusion can be considered as fusion calculation, and there are many methods to implement it, such as weighted average, Kalman filtering, Bayesian estimation, entropy theory, and fuzzy reasoning.
- **Decision making.** As the core part of unmanned driving, decision making receives sensing fusion information, learns environmental information through intelligent algorithms, and plans specific driving tasks from a global perspective so as to realize the control of unmanned vehicles. The complexity of traffic flow influences the complexity of planning tasks by means of information transmission and then determines the intelligent driving action. By constantly monitoring the vehicle movement state and the surrounding environment information, the decision making function replans tasks in real time according to the requirements of road conditions and makes the corresponding adjustment.
- **Trajectory planning.** Trajectory planning is based on the local environment information, the upper decision task, and real-time vehicle posture information to plan and determine the desired trajectory of the vehicle in the given space and time under certain kinematic constraints, including track, speed, direction, and state. The output expected speed of the planning function and information of the driving track can be imported into the lower vehicle control execution system. The trajectory planning function should be able to make reasonable planning for various task

decompositions generated by the decision making layer. The safety and comfort of the planning results are important indexes to measure the performance of the trajectory planning layer.

- **Exception handling.** Exception handling is the safety guarantee mechanism of unmanned driving system. On one hand, exception handling maintains the safe operation of the vehicle through early warning and fault-tolerant control when the mechanical parts of the vehicle come loose and the sensor parts fail under complicated road conditions. On the other hand, when some algorithm parameters are not set reasonably and the reasoning rules are not complete, the error-repair mechanism can be established to make the unmanned vehicle jump out of the endless loop and continue to complete the task to reduce the artificial intervention. In addition, it is very important to analyse, define, and describe the causes of error states, make criteria for judging action failures, study adaptive error repair algorithms, classify the causes of each error state, as well as formulate adjustment strategies accordingly. In brief, exception handling is an important part of the planning function and is necessary to improve the intelligence level of the vehicle.

3.1.2.3 Control

The core technologies of self-driving control are vertical control and lateral control of vehicles. Vertical control means driving and braking control of vehicles, while lateral control represents steering wheel angle adjustment and tire force control. With the realization of vertical and lateral automatic control, the unmanned vehicle can operate according to the given target and constraints. Therefore, from the point of view of the vehicle, unmanned driving is a combination of vertical and lateral control. In order to realize unmanned driving, the vehicle control system must acquire the detailed dynamic road information as well as the surrounding traffic conditions with highly intelligent control performance. Real-time traffic information system, highly reliable vehicle sensor, and intelligent control system are important prerequisites for the realization of unmanned driving. Vertical and lateral controls will be described in detail in the following.

- **Vertical control**. The vertical control of unmanned vehicle is the control in the direction of driving speed, that is, the automatic control of the speed and the distance between the vehicle and the front and rear vehicles or obstacles. Cruise control and emergency braking control are typical self-driving vertical control cases. Such control problems can be

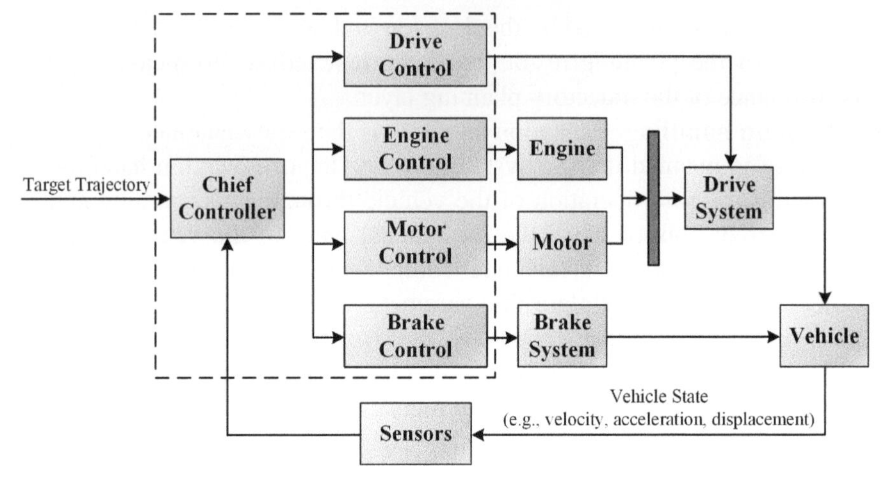

Figure 3.4 Typical structure of vertical control.

considered as the control of motor drive, engine, as well as driving and braking systems. Various drive models, operation models, and brake process models are combined with different controller algorithms to form a variety of vertical control modes. Typical structure of vertical control is shown in Figure 3.4. In addition, the slip rate control system for tire force is a key part of vertical stability control. It can control the wheel slip rate to adjust the vertical dynamic characteristics of the vehicle and prevent the excessive driving slip or braking, thus improving the stability and handling performance of the vehicle.

Intelligent control strategies, such as fuzzy control, neural network control, and receding horizon control, are widely studied and applied in vertical control. Many traditional control methods, such as proportion integral differential (PID) control and open-loop feedforward control, generally establish the approximate linear models of engine and automobile motion process to design the controller on this basis. However, the precision and adaptability of these methods are poor. Simple and accurate vehicle motion models, as well as controllers that are robust to random disturbances and adaptable to changes, are still the main focus of vertical control research. At present, the vertical control systems, such as cruise control and anti-collision control, all use vehicle sensors to obtain the information needed for control but often lack the use of V2X technology. In intelligent transportation environment, vehicles can obtain more peripheral traffic flow information through V2X communication

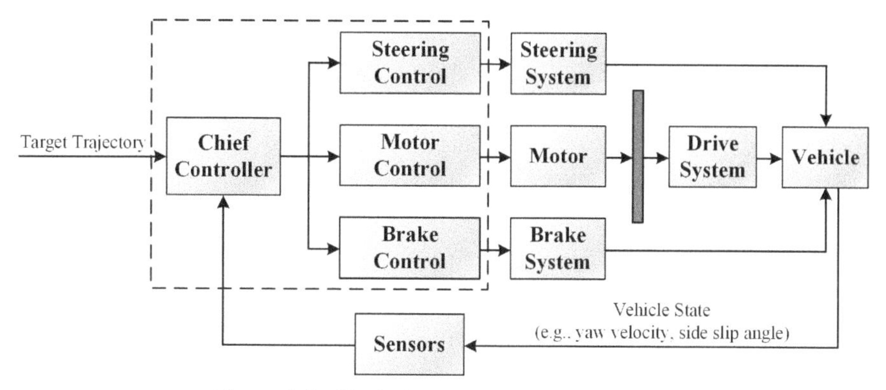

Figure 3.5 Typical structure of lateral control.

system. In the aspect of vertical control, information such as the position of the vehicle and its surrounding vehicles, as well as the current road conditions, can be used to realize predictive control so as to achieve the purpose of safe, efficient, and energy-saving driving. Lateral control refers to the control perpendicular to the direction of motion, such as steering control. The goal of lateral control is to control the vehicle to automatically maintain the desired route and to have good driving comfort and stability under different speed, load, wind resistance, and road conditions. There are two basic design methods for vehicle lateral control: one is based on driver simulation, while the other is based on the lateral motion mechanics model, which should establish more accurate vehicle lateral motion model. The typical model is called monorail model or bicycle model; that is, the characteristics of the left and right sides of the car are the same. The control target is generally the offset between the centre of the vehicle and the centre line of the road and is constrained by the comfort index. The basic structure of the lateral control system is shown in Figure 3.5. Moreover, in the traffic environment with intelligent network connection, vehicles can obtain more peripheral traffic information through environment sensing, positioning navigation, and V2X communication system for lateral control so as to facilitate the early perception of road danger.

The three core functions of unmanned driving can also be summarized as sensing, understanding, and acting. Figure 3.6 shows the data flow in driverless systems. Sensing is the function of driverless vehicle to receive environmental information, and the data it acquires will be the input of the

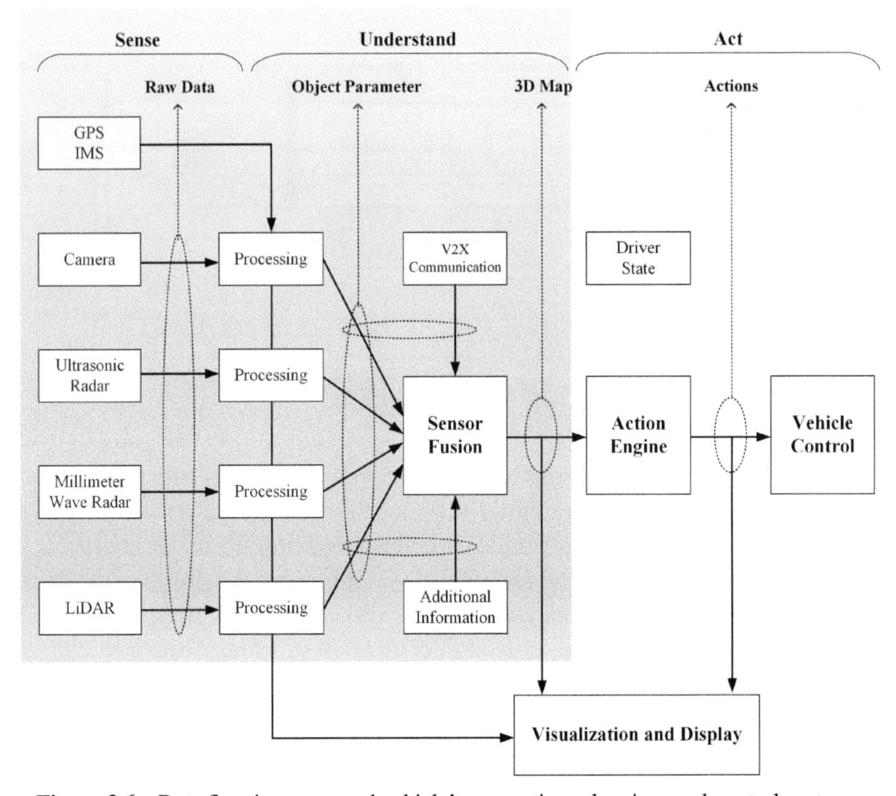

Figure 3.6 Data flow in unmanned vehicle's perception, planning, and control systems.

understanding part. The algorithms of the understanding part are responsible for guiding what the unmanned car can do. Finally, the acting part takes the result from the understanding part as the input to guide the driving of unmanned vehicles. With these three core functions, driverless cars can have the ability to obtain and extract information from the surrounding environment and adapt to the changing road conditions.

3.2 Ultrasonic Radar Security

3.2.1 Overview of Ultrasonic Radar

Vehicle ultrasonic radar system is generally composed of sensors, controllers, and monitors. Most of the ultrasonic radars in the market now adopt the principle of ultrasonic ranging. Under the control of the controller, the probe

Figure 3.7 Ultrasonic radar mounted on the rear of the vehicle.

installed at the rear (see Figure 3.7) or front of the vehicle sends ultrasonic waves and encounters the obstacles. Nowadays, there are 40, 48, and 58 KHz of commonly used probes. Generally speaking, the higher the frequency, the higher the sensitivity, but the smaller the detection angle in horizontal and vertical direction. Note that 40 KHz probe is the most widely used.

There are two common types of ultrasonic radars. The first is the ultrasonic parking assist (UPA) sensor installed on the front and rear of a car to measure obstacles. The second is auto parking assist (APA) sensor, an ultrasonic radar that is mounted on the side of a car and is used to measure the distance of obstacles on the side. The detection range of UPA ultrasonic radar is generally between 15 and 250 cm. The detection range of APA ultrasonic radar is generally between 30 and 500 cm. APA sensor's detection range is farther, and, therefore, it is more costly and powerful than UPA sensor. In the following, ultrasonic radar will be introduced in detail from three aspects, that is, detection angle, range, and sensitivity.

- **Detection angle.** The probe can only receive ultrasonic waves in a certain angle range, which is called the detection angle. It is divided into horizontal direction and vertical direction. The detection angle of ultrasonic radar is $90° - 120°$ in horizontal direction and $60° - 80°$ in vertical direction.
- **Detection coverage.** Usually, the more probes there are, the wider the coverage of ultrasonic radar is, that is, the smaller the blind area is. Nowadays, there are many kinds of reversing radars on the market, such as radars with two probes, three probes, four probes, six probes, or even eight probes. Reversing radars with 2–4 probes are usually mounted on the rear bumper of a car, while those with 6–8 probes are usually separately at the front and rear of a car.

- **Detection sensitivity.** The detection sensitivity of ultrasonic radar depends not only on the structure and material of the probe itself but also on the intensity of the reflected echo, which is related to the propagation characteristics of ultrasonic wave and can be summarized as the following points.
    1. Due to the attenuation of ultrasonic wave transmission in air, for the same reflecting surface and the same angle, farther distance will result in greater attenuation of reflected ultrasonic wave, making it harder to be detected.
    2. The larger the reflecting surface area of the obstacle is, the stronger the reflected wave is, and the farther the detection distance is. Under the following circumstances, even if the obstacle is close to the ultrasonic radar, it may not be detected. For example, the reflecting surface of the obstacle is small, and the reflecting surface of the obstacle is large but deviates from the direction of the ultrasonic sensor. In these cases, the ultrasonic radar may receive little or no reflection.
    3. Obstacles on the centre line of the sensor have the strongest reflection wave and the farthest detection distance and vice versa.
    4. The barrier will absorb some of the ultrasonic wave and reflect back only part of it. The amount of absorption or reflection is related to the material and surface of the obstacle. Generally speaking, loose and porous object surface is easier to absorb ultrasonic wave. Therefore, its reflection efficiency is lower, and it is not easy to be detected. On the contrary, if the material of the obstacle is harder, the reflection wave will be stronger.
    5. Environmental temperature, air humidity, air pressure, and other factors will affect the intensity of reflected echo. In short, heavy rain, heavy snow, or undercooling, overheating, and wet weather can all affect the ultrasonic radar's detection performance.

According to the above characteristics, it is not easy to detect objects with sharp-angle reflecting surface and multi-angle reflecting surface, low position water pipes, wire mesh, rope and other small objects, as well as cotton or other surface that can easily absorb ultrasonic wave, etc. Besides, because the ultrasonic frequency used is generally 40 KHz, if there are other interference sources of similar frequency nearby, it is easy to cause false alarm or slow reaction. Common interference sources include spray paint, refrigerant filling equipment, pneumatic tools, high-power exhaust fan, and so on. Specifically, compared with other sensors, ultrasonic radar has the following advantages and disadvantages.

- **Advantages of ultrasonic radar**:
  1. The energy consumption of ultrasonic wave is small, the propagation distance is relatively long, and its penetrability is strong.
  2. The principle of ultrasonic radar ranging is simple. As described in the following chapter, the distance can be directly derived from propagation speed and propagation time.
  3. Low cost: LiDAR costs tens of thousands of dollars, while ultrasonic radar only costs a few dollars with its mature and stable technology.
- **Disadvantages of ultrasonic radar**:
  1. Sensitive to temperature: The ranging principle of ultrasonic radar is similar to that of LiDAR and millimeter wave radar. The difference is that the wave velocity of LiDAR and millimeter wave radar is the speed of light, while the wave velocity of ultrasonic radar is related to temperature. The approximate relationship of ultrasonic wave's velocity (C) and the temperature can be represented as follows:
  $$C = C_0 + 0.607 \cdot T, \tag{3.1}$$
  where C_0 is the velocity of the sound wave at 0 °C, and T is temperature, measured in °C. For obstacles in the same relative position, the measured distance is different at various temperatures. For the unmanned driving system with high requirement for sensor precision, the ranging of ultrasonic radar can be conservatively calculated, and the temperature information can also be introduced into the unmanned driving system to improve the measurement accuracy.
  2. The location of the obstacle cannot be accurately described. Ultrasonic radar will return a value of detection range when in operation. If the obstacles in different positions return the same detection distance, the location of the obstacles cannot be determined through the information of a single radar under the condition that only the detection distance is known.
  3. When the car is running at high speed, the distance measurement of ultrasonic radar has some limitations. Ultrasonic ranging cannot

keep up with the real-time change of vehicle speed, which affects the measurement accuracy.

4. The scattering angle of ultrasonic wave is large and the directionality is poor. When measuring the distant target, the echo signal will be weak, which will also affect the measurement accuracy. However, ultrasonic ranging sensor has great advantages in short distance measurement.

3.2.2 Basic Principle of Ultrasonic Radar

Ultrasonic sensors detect objects by emitting ultrasonic pulses and measure the time it takes for the echo pulse to reflect back from an obstacle, as Figure 3.8 shows. The distance d to the nearest obstacle is calculated according to the propagation time of the first received echo pulse, which can be represented as follows:

$$d = 0.5 * t_e * c, \tag{3.2}$$

where t_e is the propagation time of the ultrasonic echo and c is the speed of sound in air (about 340 m/s).

The acoustic part of the ultrasonic radar is a piezoelectric transducer, and the operation of ultrasonic sensor is based on piezoelectric effects. The piezoelectrics are electrically polarized by the external mechanical force, and the opposite bound charge appears on the surface of the two ends of the piezoelectrics. The charge density is proportional to the external mechanical force. This phenomenon is called positive piezoelectric effect. The piezoelectrics are deformed by the action of the external electric field, and their shape

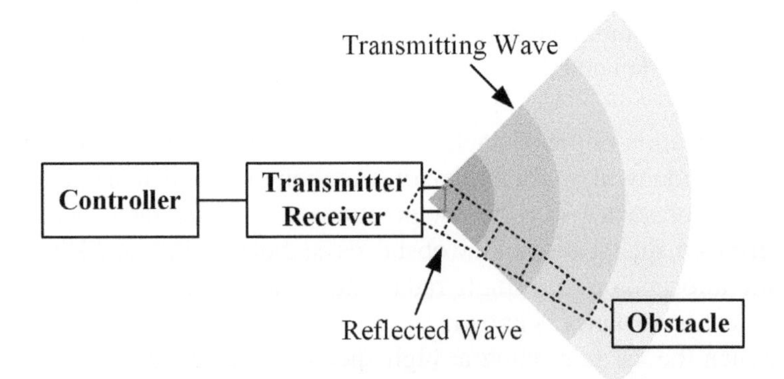

Figure 3.8 Working principle of ultrasonic radar.

variable is proportional to the intensity of the external electric field. This phenomenon is called the inverse piezoelectric effect. Solids with positive piezoelectric effects also have inverse piezoelectric effects and vice versa. Positive piezoelectric effect and inverse piezoelectric effect are collectively called piezoelectric effect. Whether the crystal has piezoelectric effect is determined by the symmetry of its structure.

When the digital transmission signal from electronic control unit (ECU) is received by the sensor, the circuit stimulates the diaphragm to vibrate with a square wave at a resonant frequency of 40–50 KHz and emits ultrasonic waves. Once stopped, the diaphragm can vibrate again through the echoes reflected back from the obstacle, which are converted by the piezoelectric crystals into analog signals and are then amplified, filtered, digitized, as well as compared with thresholds to determine the arrival of the echoes. Finally, the echo transmission time graph is transmitted to the ECU to further calculate the distance. Ultrasonic transducers typically operate at frequencies between 40 and 50 KHz, which has proven to be the best trade-off between good acoustic performance in terms of sensitivity and range and high robustness against ambient noise around the sensor.

3.2.3 Attack Technique Analysis of Ultrasonic Radar

Based on the above knowledge, if an attack system can be designed to produce the ultrasonic wave with the same frequency as the on-board ultrasonic radar, the jamming attack and spoofing attack can be launched by observing the sensor's reaction and the unmanned driving system's reaction.

The function of ultrasonic radar on driverless vehicle is to detect obstacles. After emitting the ultrasonic wave, the distance between vehicle and obstacle is judged according to the time of receiving echo. For example, 12 ultrasonic sensors are installed on Tesla Model S to detect obstacles around the vehicle. In addition, an alarm will be issued when obstacles are detected. The obstacles and distance warnings will be displayed at corresponding areas on the instrument panel.

There are two ways to affect ultrasonic radar:

- emit ultrasonic noise and increase the signal-to-noise ratio of ultrasonic sensor to affect the judgement of ultrasonic radar, thus leading to the error display of distance information on the instrument panel;
- emit ultrasonic waves of the same frequency as the equipped ultrasonic radar, causing the ultrasonic radar to make wrong judgement.

These ways can be achieved by a low-cost jamming device that sends corresponding signal in front of the vehicle's ultrasonic sensors. Under manual driving, the driver will be misled by the error message displayed on the dashboard, while under unmanned driving, the unmanned driving system will also be misled, leading to false alarm. In the second case, if the signal is sent at the same frequency for jamming, the attack time is important. Note that only the first received ultrasonic wave is valid. In addition, using acoustic absorbent materials and pertinent algorithms can also realize acoustic cancellation.

3.2.3.1 Jamming Attack

The principle of jamming attack is to generate ultrasonic noise and cause continuous vibration of the diaphragm on the sensor, as shown in Figure 3.9. Therefore, the measurement cannot be carried out normally, and the ultrasonic radar cannot accurately detect obstacles.

The researchers found that the performance of the ultrasonic radars decreased under two conditions. In the first condition, the ultrasonic transmitter in the ultrasonic working frequency near the vehicle will reduce the signal-to-noise ratio, leading to wrong measurement of the on-board ultrasonic sensor. In the second condition, any rain, snow, ice, or dirt on the sensor diaphragm can form a sound bridge with the buffer. These vulnerabilities make physical attacks on ultrasonic sensors possible.

In order to simulate the external noise source, ultrasonic transducer is a good choice. The function of an ultrasonic transducer is to convert incoming electrical signals into ultrasonic waves for transmission, while consuming

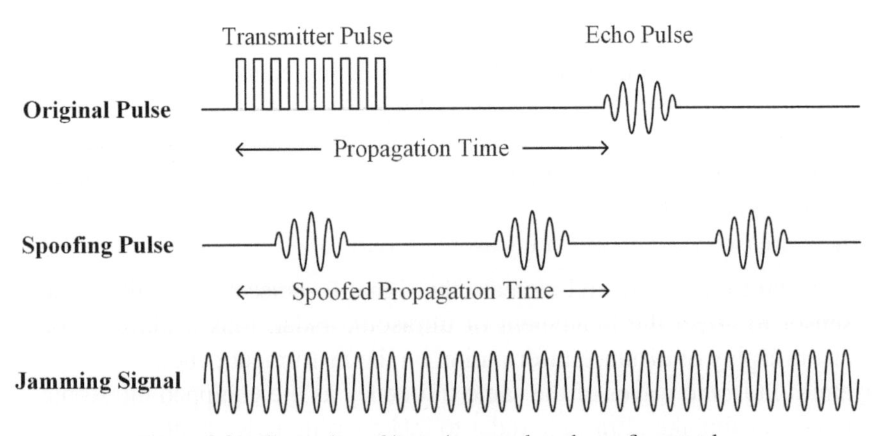

Figure 3.9 Illustration of jamming attack and spoofing attack.

a small amount of power on its own. The transducer consists of housing, matching layer, piezoelectric ceramic disc transducer, backing, lead cable, and Cymbal array receiver. It has an extremely high sound pressure level, excellent frequency performance, and controllability.

In a reverse study, the researchers found that Tesla's ultrasonic radars emit waves of 40 KHz, such as when shaking a key string or braking a large truck. However, in daily life, such ultrasound does not last long and is not very intense. The researchers tried a jamming attack on Tesla's ultrasonic radars, playing the same wavelength at a higher intensity so that the ultrasonic radars could not pick up their own signals and thus could not accurately measure the distance of objects around them.

3.2.3.2 Spoofing Attack

The spoofing attack is similar to the jamming attack, and its purpose is to spoof the ultrasonic radar. The consequence of such attack is that the ultrasonic radar does not work properly or is manipulated. Such consequences can be exploited by attackers to cause serious accidents.

Through the signal analyser to further crack the ultrasonic signal, ultrasonic structure has been mastered by researchers. They tried to trick the sensors with a signal transmitter. If the spurious ultrasonic pulse can be identified as an echo from an obstacle and reaches the sensor before the real echo, the sensor reading will deviate from the real value. By setting the trigger time of the fake pulse, an attacker can forge the measured distance of the ultrasonic sensor.

The specific experiments of jamming attack and spoofing attack can be found in Section 3.6.

3.3 Millimeter Wave Radar Security

3.3.1 Overview of Millimeter Wave Radar

Millimeter wave radar is a key component of self-driving vehicles. Its ranging principle is similar to the aforementioned ultrasonic radar, that is, the radar wave is first sent out, then the echo is received, and the position of the target is measured according to the time difference between the transceiver and the receiver.

The wavelength of millimeter wave is between centimetre wave and light wave. Compared with the centimetre wave seeker, the millimeter wave seeker has the characteristics of small volume, light weight, and high spatial

Table 3.2 Parameter Comparison of Millimeter Wave Radar.

	Short- or Medium-Range Millimeter Wave Radar	Long-Range Millimeter Wave Radar
Range	30 or 120 m	280 m
Support Speed	150 km/h	250 km/h
Accuracy	Centimetre-level	0.5 m
Application	Environment monitoring	Adaptive cruise control

resolution. Compared with infrared, laser, TV, and other optical seeker, millimeter wave seeker has strong ability to penetrate fog, smoke, and dust. In addition, the anti-interference and anti-stealth abilities of millimeter wave seeker are better than other microwave seekers. Note that millimeter wave radar has low accuracy and small viewing angle; so it generally requires a combination of multiple radars to achieve better results.

Millimeter wave radars are available in 24, 60, 77, and 79 GHz. The main available frequency bands are 24 GHz and 77 GHz, which are used for medium and long distance measurement, respectively. Long-range millimeter wave radar has improved speed measurement and ranging accuracy as well as a smaller volume, making it easier to be deployed on vehicles. As shown in Table 3.2, the long-range radar has a wider detection range and can be adapted to faster vehicles, but the corresponding detection accuracy is reduced.

Vehicle-mounted millimeter wave radar transmits millimeter wave outward through the antenna and then receives the target reflection signal. It can quickly and accurately acquire the environment information around the vehicle body (e.g., the relative distance between the vehicle and other objects, relative speed, angle, direction of motion, etc.) and then track and classify the target according to the detected object information. After that, it combines and processes the dynamic information for data fusion. Finally, the driver is informed or warned by sound, light, and touch, or active intervention by unmanned driving system is made in time to ensure the safety and comfort of the driving process and reduce the probability of accidents.

Millimeter wave radars can be divided into pulse mode and continuous wave mode according to different measurement principles, in which continuous wave includes frequency shift keying (FSK), phase shift keying (PSK), constant frequency continuous wave (CFCW), frequency modulated continuous wave (FMCW), and so on. Each of these modes has its own characteristics, and a brief introduction of them is shown below.

- **Pulse mode**. The millimeter wave radar using pulse mode needs to transmit high-power pulse signal in a short time to control the oscillator.

At the same time, the echo signal should be strictly isolated from the transmitted signal before amplification.

- **Continuous wave mode**. The millimeter wave radar with FMCW ranging is simple in structure and small in volume. Its biggest advantage is that the relative distance and velocity of the target can be obtained simultaneously. When the continuous frequency modulation signal transmitted by radar meets the target in front, the echo with a certain delay relative to the transmitted signal will be generated, and then the mixing process will be carried out through the radar mixer, and the result after mixing is related to the relative distance and relative velocity of the target.

The distance, angle, and relative velocity between the radar and the object can be measured by mounting the millimeter wave radar on the vehicle. Millimeter wave radar is mainly used in high-end cars and can be adopted to realize the adaptive cruise control (ACC), forward collision warning (FCW), blind spot detection, auxiliary parking, lane change assistant, and other advanced driving assistance system (ADAS) features. Among them, the 24-GHz radar system mainly realizes short-range detection, while the 77-GHz radar system mainly realizes long-range detection.

3.3.2 Basic Principle of Millimeter Wave Radar

Millimeter wave radar mainly works in microwave frequency band. It detects the target by transmitting electromagnetic wave and receiving reflected echo. The millimeter wave radar has the concept of beam, which means that the electromagnetic wave emitted is a conical beam, rather than a laser line. This is because the antennas in this band work mainly by electromagnetic radiation, rather than optical particle emission. At this point, millimeter wave radar and ultrasonic radar are the same. This kind of beam also leads to millimeter wave radar's advantages and disadvantages. Millimeter wave radar has the advantage of good reliability because of the large reflector, while its disadvantage is low discrimination. Note that millimeter wave radar can detect the distance, velocity, and azimuth of the target.

Millimeter wave radar mainly has two working modes: pulse mode and continuous wave mode. However, the operating range of millimeter wave radar is not too far, and the interval between the echo and the emission wave is very short; so it is not suitable to use the simple pulse mode. The on-board millimeter wave radar usually adopts FMCW radar system with simple structure and low cost, which is suitable for short-range detection. The radar

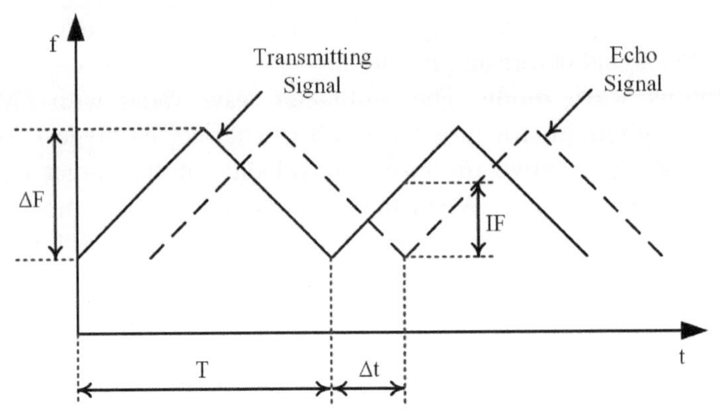

Figure 3.10 Comparison of transmitting signal and echo signal.

antenna sends out a series of continuous FM millimeter waves, and the frequency varies with time according to the regulation of modulation voltage, which is generally a triangular wave. The reflected and transmitted signals have the same waveform but with a delay time Δt. Taking the triangular wave signal as an example, the contrast diagram between the transmitted signal and the returned signal is shown in Figure 3.10.

3.3.2.1 Ranging Principle

The frequency modulator of radar transmits continuous wave signal through antenna. When the transmitting signal meets the target, the echo signal will be generated by the reflection of the target. Compared with the echo signal, the transmitting signal has the same shape, but there is a difference in time. The following formula is used for ranging stationary objects:

$$R = \frac{C \cdot \Delta t}{2}, \tag{3.3}$$

where R is the distance to the target object and C represents the speed of light. The transmitting signal and the echo signal have the same shape; so the following equation can be obtained according to the relation of trigonometric function:

$$\frac{\Delta t}{\mathrm{IF}} = \frac{T/2}{\Delta F}, \tag{3.4}$$

in which T is the period of the transmitted signal, ΔF is the frequency modulation bandwidth, and IF is the intermediate frequency of transmitted signal mixed with echo signal. In this way, the relation between target distance R

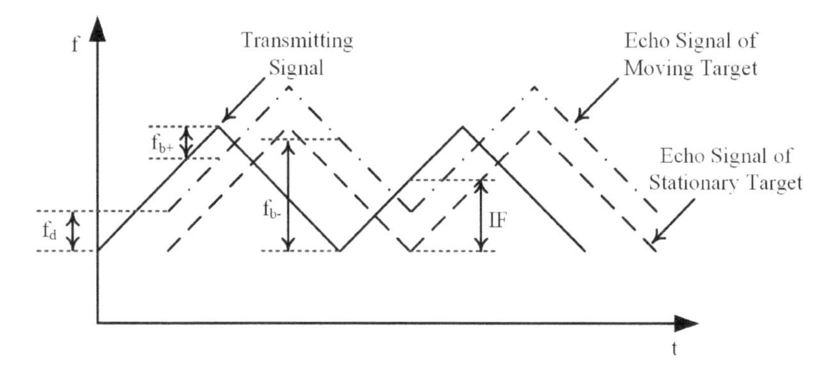

Figure 3.11 Doppler shift occurs between the frequency of echo signal and the transmitted signal.

and IF signal can be denoted as follows:

$$R = \frac{C \cdot T \cdot \mathrm{IF}}{4 \cdot \Delta F}. \tag{3.5}$$

3.3.2.2 Velocity Measuring Principle

When there is a relative motion between the target and the radar signal emitter, in addition to the time difference between the reflected echo signal and the transmitted one, the Doppler shift will occur between the frequency of the echo signal and the transmitted signal. The frequency of an intermediate frequency signal in the ascending and descending phases can be expressed as follows:

$$f_{b+} = \mathrm{IF} - f_d, \tag{3.6}$$

$$f_{b-} = \mathrm{IF} + f_d, \tag{3.7}$$

where f_d is the Doppler shift between transmitted signal and echo signal and can be represented as follows:

$$f_d = \frac{f_{b-} - f_{b+}}{2}. \tag{3.8}$$

According to Doppler principle, the velocity measurement of relative moving objects are based on the following formula:

$$V = \frac{(f_{b-} - f_{b+}) \cdot C}{4 \cdot f_0}, \tag{3.9}$$

where f_0 is the central frequency of the emitted wave.

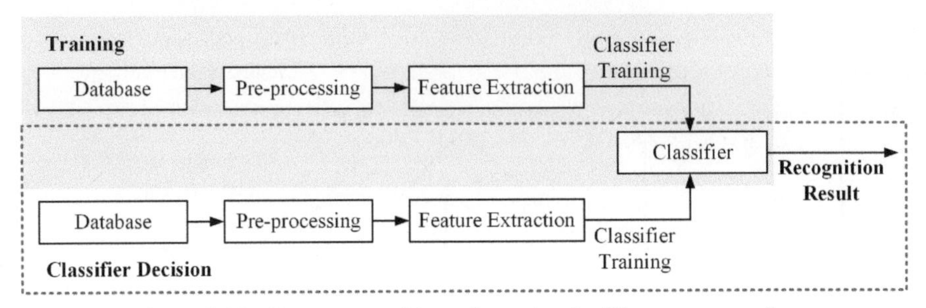

Figure 3.12 Target recognition schematic of millimeter wave radar.

3.3.2.3 Target Recognition

The basic principle of target recognition of millimeter wave radar is to estimate the physical characteristics such as size and shape by multidimensional spatial transformation using target feature information such as amplitude, phase, spectrum, and polarization in radar echo.

According to the identification function determined by a large number of training samples, the recognition judgement is carried out in the classifier, including target recognition preprocessing, feature signal extraction, feature space transformation, pattern classifier, and sample learning. The feature signal extraction refers to the millimeter wave radar's acquisition and extraction of various information generated by the interaction between the electromagnetic wave and the target, including the radar scattering cross-sectional area and other characteristic parameters. The commonly used feature parameters include the structural features of the target, the dynamic characteristics of the target, and the echo waveform characteristics. The purpose of feature space transformation is to change the original data distribution structure, compress the dimension of feature space, and remove redundant features. The commonly used feature transformation techniques include Karhunen–Loeve (K-L) transform, Walsh transform, and so on. The millimeter wave radar's target recognition schematic is shown in Figure 3.12.

3.3.3 Attack Technique Analysis of Millimeter Wave Radar

The millimeter wave radar has strong penetration. Its ranging accuracy is less affected by rain, snow, and fog weather. Millimeter wave radar is the main research direction of automobile anti-collision radar because of its small size. At present, the main working frequency bands of automobile millimeter wave radar are 24, 36, 60, and 77 GHz. The millimeter wave radar assembled by

Tesla operates at about 77 GHz. This ultra-high frequency technology was once used as a secret technology for the U.S. military. Millimeter wave radar is installed in the front of Tesla car to detect long-distance obstacles and can identify obstacles up to 150 m away. The 77-GHz ultra-high frequency has exceeded the range that the general instrument can parse. Therefore, the characteristics of millimeter wave radar need further analysis. On one hand, attackers can analyse through manufacturers and search for vehicle information. On the other hand, these characteristics can also be obtained by directly observing millimeter wave radar's spectrum and waveform.

3.3.3.1 Jamming Attack

Compared with ultrasonic radar, millimeter wave radar is much more difficult to be cracked. One of the biggest reasons is the high cost of the attack. The cost of equipment that interferes with millimeter wave radar is almost equivalent to the price of three Tesla Model S cars. Through the interference of electromagnetic wave, the millimeter wave radar can think there is a car in front of no car, and this information will be displayed on the dashboard synchronously. If the distance is close enough, the alarm component will also raise a warning, reminding that the vehicle may brake automatically in autopilot mode. Similarly, jamming attack on millimeter wave radar could enable the unmanned driving system to assume that there is no car ahead and continue driving when there is a car in front.

Knowing the characteristics of millimeter wave radar, the signal generator can be used to generate electromagnetic wave in the same frequency band to jam the millimeter wave radar system and attack it directly. Jamming attack can make the detected object disappear from the autopilot system. Therefore, an attacker could make an unmanned vehicle at high speed completely ignore obstacles in front of it or even make the car brake suddenly in an empty environment. In theory, such attacks can take place tens of meters away. However, the beam of millimeter wave transmitter is concentrated. In an actual situation, to perfectly hit the millimeter wave radar mounted on a moving vehicle requires high accuracy.

3.3.3.2 Spoofing Attack

Spoofing attacks can change the measured distance of the target. The results can be obtained by the same modulation signal as the vehicle radar. Periodic distance changes shown in Tesla car can be observed by adjusting the slope back and forth in the higher value range on the signal generator.

3.3.3.3 Relay Attack

The principle of the relay attack is to relay the received signal to the transmitter in the harmonic mixer and then send it back to the millimeter wave radar to simulate further targets. In this attack, the relay signal is very close to the real one, and it is less likely to be suspected; so the probability of success is high.

The detailed experiment of jamming attack against Tesla Model S can be found in Section 3.6.

3.4 Hi-Definition Camera Security

3.4.1 Overview of Hi-Definition Camera

Hi-definition camera is a necessary sensor for self-driving cars. Unlike radar, the camera has no penetration and requires light for illumination. Most of the data used for unmanned driving is obtained by image recognition of the camera. However, the camera is also the most easily disturbed self-driving sensor. Once the obtained image has errors, the final recognition results will be greatly affected. The advantage of hi-definition camera is that it is of low cost, the current vision recognition scheme is relatively mature, and there are many camera schemes available for self-driving cars. Many functions can be realized by camera, such as ACC, lane departure warning (LDW), lane keeping assist (LKA), FCW, autonomous emergency braking (AEB), traffic sign recognition (TSR), automatic parking (AP), pedestrian detection system (PDS), driver monitor status (DMS), surround view cameras (SVC), and so on, while some functions can only be realized by camera.

Cameras cost less than other sensors and are easier to be applied in unmanned driving. According to the requirements of different ADAS functions, the camera can be installed in different positions. The front-view camera is usually a wide-angle lens, which is installed on the rearview mirror or on the front windshield in order to realize the long effective distance. Through the development and optimization of algorithm, a single forward view camera can achieve multiple functions, such as driving record, LDW, FCW, PDS, and so on. For example, Tesla Autopilot 2.0 hardware system contains eight cameras, including three front view cameras, i.e., normal, long-focus, and wide-angle cameras. Using three cameras can cover longer distance and wider range of vision, and the detection accuracy and security can be greatly improved. In the future, multi-camera system will become a trend. In addition, the rearview mirror has a limited range and a blind area.

The side-view cameras on both sides of the vehicle can basically cover the blind area. The side-view wide-angle camera can also replace the rearview mirror, which can not only reduce the wind resistance but also obtain a larger and wider perspective to avoid accidents in dangerous blind areas. Japan has also revised regulations to allow vehicles without rearview mirrors to drive on the road, encouraging the use of side-view cameras to replace rearview mirrors, and the National Highway Traffic Safety Administration (NHTSA) has also promised to amend regulations to remove restrictions on vehicles without rearview mirrors.

Hi-definition cameras are available in a wide variety of models and types, which can be simply divided into monocular camera, binocular stereo camera, and panoramic camera.

- **Monocular camera**. Environmental imaging of unmanned vehicles is the application of machine vision in vehicles, which needs to meet the requirements of driving environment and driving condition. Changes in weather, vehicle speed, vehicle trajectory, random disturbance, and camera installation position will all affect the on-board vision system. Self-driving cars not only need high frame rate in image output speed but also have high requirement in image quality. Monocular camera is a camera that only uses a set of optical system and solid state imaging device to continuously output images, which can realize real-time adjustment of optical integration time and automatic white balance.
  The general ranging principle of monocular camera is to first identify the target through image matching and then estimate the distance of it through the size in the image, which requires accurate identification of the target before estimation. To achieve this, it is necessary to establish and maintain a large sample feature database that contains all feature data of the target to be identified. If there is no characteristic data of the target to be identified, the system will be unable to identify objects and obstacles and, thus, cannot accurately estimate the distance to these targets. The application scenarios of monocular camera include ACC, LDW, LKA, FCW, AEB, TSR, AP, PDS, and DMS. The advantages and disadvantages of monocular cameras are shown in Table 3.3.
- **Binocular stereo camera**. The binocular cameras are capable of stereoscopic imaging of objects in the field of view. Its design is based on the study of human visual system. It uses binocular stereo image processing to get the three-dimensional information and the depth map. Then, the scene in the three-dimensional space can be obtained after

Table 3.3 Advantages and Disadvantages of Monocular Camera.

Advantages	Low cost
	Low requirement for computing resources
	Relatively simple system structure
Disadvantages	Require continuous updating of large sample database
	Cannot identify non-standard object
	Low accuracy of distance measurement

Table 3.4 Advantages and Disadvantages of Binocular Camera.

Advantages	Cost is lower than LiDAR and many other schemes
	Obstacles can be measured directly
	High precision and direct use of parallax measurement
	No need to maintain the sample database
Disadvantages	Hardware cost is higher than monocular camera
	High requirement for computing resources
	Relatively big size
	Difficult to registration

further processing, and the reconstruction of two-dimensional image to three-dimensional image can be realized. However, in unmanned driving applications, the two imaging systems of binocular cameras may not be able to extract features perfectly.

Binocular camera is based on the principle of binocular triangulation ranging, which directly measures the distance of the object in front by calculating the parallax of two images, without judging what kind of obstacle appears in front. The human eye can perceive distances because of the parallax between two eyes on the same object. The farther the object is, the smaller the parallax will be and vice versa. The application scenarios of binocular camera is similar to monocular camera, including ACC, LDW, LKA, FCW, AEB, TSR, AP, PDS, and DMS. The advantages and disadvantages of monocular cameras are listed in Table 3.4.

- **Panoramic camera**. The panoramic camera represented by the Ladybug camera of Canadian Point Grey company is used for street view imaging. It is composed of six same cameras for simultaneous imaging. Then six images will be corrected and spliced to obtain panoramic images of simultaneous imaging.

  An unmanned car equipped with a panoramic camera can simultaneously obtain panoramic images of the surrounding environment and conduct processing as well as target recognition. Moreover, the monocular camera using fish-eye lens can also present panoramic images. Although the distortion of the original image is large, the calculation task is relatively small and the price is low. As a result, such cameras are beginning to attract attention in the field of unmanned driving. The application scenarios of panoramic camera includes AP and SVC.

Data from ultrasonic radar, millimeter wave radar, LiDAR, and many other sensors are insufficient for safe unmanned driving, especially on highways and city streets where many rules and regulations are applied. For autonomous driving systems with human drivers, the necessary information needs to be obtained from road signs and lane vision. The vehicle-mounted camera system can handle the visual recognition of the driving environment, including lane lines, traffic signs, lights, vehicles, pedestrians, and other obstacles. When the camera is combined with other sensors, driving behaviour and routes can be planned in a better and safer way; so multi-sensor fusion technology is also important.

3.4.2 Basic Principle of Hi-Definition Camera

Monocular and binocular cameras are two main categories of on-board cameras. Monocular camera is generally equipped with charge-coupled device (CCD), while binocular camera is generally with complementary metal oxide semiconductor (CMOS). Compared with CCD, the image quality of CMOS is slightly lower. However, CMOS costs less and is more energy-efficient. It is widely used in the field of on-board cameras with low pixel requirements. In general, CMOS has two important advantages over CCD.

- CMOS photoelectric sensor can take out the electric signal while collecting the optical signal and can also process the image information of each unit at the same time. Its speed is much faster than CCD. The image acquisition speed of high-performance CMOS camera can reach up to 5000 frames/second.
- When the vehicle is running at high speed, the light condition changes violently and frequently. CMOS can quickly identify surrounding objects even in environments with large differences in brightness.

As shown in Figure 3.13, hi-definition camera utilizes CCD or CMOS device and filters to collect optical data to generate images in the camera module, which are then sent to the micro control unit (MCU) for further processing and calculation. The identification results will be sent from the

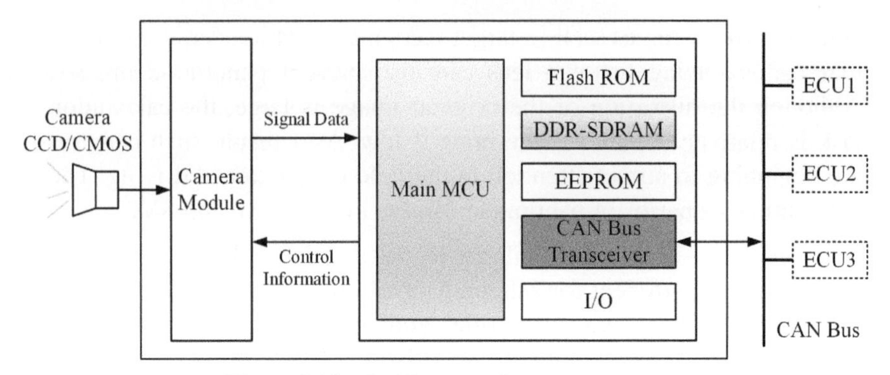

Figure 3.13 Architecture of camera system.

controller area network (CAN) bus to each ECU in the unmanned driving system. Then, the integrated processor makes driving decisions and sends commands to the actuator. Some unmanned driving systems also provide video output on vehicle's centre screen for reference.

3.4.3 Attack Technique Analysis of Hi-Definition Camera

Hi-definition camera can also be exploited by attackers. This is not to say that the attacker controls the camera by hacking into the system because, for an unmanned car, what is seen through the camera may not have any impact on it. The point is to stop the camera from working properly.

LiDAR and camera are a good match for most of today's driverless cars, working together to help with three-dimensional modelling and high-precision mapping. The former provides accurate point cloud data, while the latter outputs noisy two-dimensional image information that is often used to colour the data.

The unmanned car is able to drive on the road because the laser rangefinder tells it what the road is, and the camera information is then used to match colour to tell the vehicle what is in front of it. In contrast, considering from the view of interference, it is a common idea to make the image output by the camera appear more lines and noise points. For example, there are a variety of anti-monitoring devices on the market. In fact, their principle is to make the output quality of surveillance camera screen poor by means of electromagnetic wave interference, high–low-frequency electromagnetic wave interference, reflection interference, and so on. In particular, if the colour of the road is used to create all the obstacles or if the obstacles are merged with the background, a certain deception effect can be achieved.

3.4.3.1 Blinding Attack

The method of blinding attack is using the light source to direct the camera or to direct the calibration board so that the reflection is directed to the camera. The final effect depends on the distance between the light source and the camera as well as the intensity of the light source. At a distance of 50 cm from the camera, 200-MW infrared ray can blind the camera for 40 seconds. If the distance increases, the effect decreases. If the light source is too strong, it will cause the camera to burn out directly.

Attacking at a distance of less than 50 cm may not be possible in real life. With the lengthening of the distance, blinding attack can be caused by increasing the intensity of the light source. As long as the blindness lasts 2–3 seconds, it is likely to have serious consequences for an unmanned car. From the view of hardware, the camera itself has a refresh frequency. If it is high enough, it will reduce the effect of the blinding attack.

3.5 LiDAR Security

3.5.1 Overview of LiDAR

LiDAR, also known as optical radar, is a ranging technology. It uses the laser light wave to transmit the detection signal to the target and then compares the signal it receives with the transmitted one so as to obtain the target's position (e.g., distance, azimuth, and height), motion state (e.g., speed and attitude), and other information. In this way, the detection, tracking, and recognition of the target can be realized.

LiDAR technology is one of the key technologies of driverless vehicles. Because of its advantages of high ranging accuracy, strong directionality, fast response, and not being affected by ground clutter, it can effectively provide the needed information for vehicle's decision making and control system and has become the most effective sensing scheme for unmanned driving. However, LiDAR is expensive and very vulnerable to environmental factors, making the detection results in bad weather greatly affected.

LiDAR can be used in both airborne and terrestrial applications. Airborne LiDAR is an airborne laser detection and ranging system installed on aircraft to measure three-dimensional coordinates of ground objects. In the 1970s, developed by National Aeronautics and Space Administration (NASA), LiDAR surveying and mapping technology began to progress. In addition to military applications, LiDAR is also rapidly expanding into the civilian market. Among them, unmanned driving is one of the hottest applications.

Defense Advanced Research Projects Agency (DARPA) holds the driverless car challenge every year. At the DARPA challenge in 2007, six of the seven teams used LiDAR designed by Velodyne company. This also drew the attention of Google, which is preparing to develop driverless cars. Google launched its driverless car project in 2009, using Velodyne's LiDAR in its prototype. In recent years, the driverless car market is booming. As Google, Baidu, Uber, and other mainstream driverless car research and development institutions have taken LiDAR as one of the main sensors, the combination with image recognition and other technologies can be used to realize the vehicle's judgement of road conditions. Traditional automakers including Volkswagen, Nissan, and Toyota are developing and testing driverless car technology, and they have also adopted LiDAR.

The advantages of LiDAR are its accuracy, speed, and efficiency. It is a kind of sensor for accurately obtaining three-dimensional position information. Its function in unmanned vehicles is equivalent to the human eye. It can determine position, size, external appearance, and even material of objects.

LiDAR determines the distance by measuring the time difference and phase difference of the laser signal and measures the angle by horizontal rotation scan or phase control scan. It establishes a two-dimensional polar coordinate system based on the aforementioned two kinds of data and then gets the height information of the third dimension by obtaining signals of different pitch angles. A high frequency laser can obtain a large amount of position point information (point cloud) in 1 second and carry out three-dimensional modelling based on this information. In addition to obtaining location information, LiDAR can preliminarily distinguish different materials by the reflectivity of laser signal. In general, LiDAR plays two core roles in unmanned vehicles as follows.

- LiDAR can realize three-dimensional modelling for environment awareness through laser scanning. By comparing the environment information of the previous frame with that of the next frame through relevant algorithms, it is easier to detect the surrounding vehicles and pedestrians.
- LiDAR can achieve visual simultaneous localization and mapping (SLAM). By obtaining the global map in real time and comparing with the features in the high-precision map, the navigation and positioning accuracy of the vehicle can be enhanced.

While LiDAR offers opportunities for unmanned driving, it also brings many challenges, which can be summarized below.

- The working principle of LiDAR is based on measurements of the return time of laser pulses; so surfaces with high reflectivity can cause problems. If the surface's reflectivity is very high, light will scatter away from the sensor, and the point cloud information in this area will be incomplete.
- The environment can also affect the performance of LiDAR. For example, fog and heavy rain can weaken the laser pulses and affect the LiDAR.
- The refresh rate of the LiDAR system is limited by the rotation speed of complex optical devices. The fastest rotation rate of the LiDAR system is about 10 Hz, which limits the refresh rate of the data stream.
- High cost is a great challenge that LiDAR needs to overcome. Although the cost of LiDAR technology has fallen dramatically since its introduction, its high price remains an important barrier to its widespread adoption.
- Although LiDAR can be seen as a component of computer vision, point clouds are based entirely on geometry. The current LiDAR system cannot distinguish between paper bags and rocks, which should be considered by on-board sensors when they try to avoid obstacles.

3.5.2 Basic Principle of LiDAR

The LiDAR consists of three parts: transmitting system, receiving system, and information processing system. The laser device turns the electrical pulse into the optical pulse and then the optical receiver reduces the reflected optical pulse from the target into the electrical pulse. Finally, a series of algorithms are used to obtain the target position, motion state, and shape so as to detect, identify, and track the target. According to different technical principles, the LiDAR can be divided into three types: mechanical LiDAR, solid-state LiDAR, and hybrid solid-state LiDAR. These different types of LiDARs will all be introduced in the following.

- **Mechanical LiDAR** changes the scanning direction in a mechanical way, that is, the laser beams are arranged vertically and rotated around the axis. Each laser beam scans a plane and presents three-dimensional graphics after being vertically superposed. The technical principle of multi-line stacking of mechanical LiDAR makes the imaging effect closely related to the number of laser lines. The more lines there are, the more planes can be scanned, the more information can be obtained from the target, and the better the imaging effect will be. Because of

Table 3.5 Advantages and Disadvantages of Mechanical LiDAR.

Advantages	Mature technical scheme
	Wide fields of view
	Support for remote detection
	High measurement accuracy under high number of lines
Disadvantages	Large in size
	High cost
	Vulnerable to damage while driving

Table 3.6 Advantages and Disadvantages of Solid-State LiDAR.

	OPA Solid-State LiDAR	**Flash Solid-State LiDAR**
Advantages	Fast scanning speed	No moving parts
	High precision	Mature transmitter scheme
	Small in size	Low cost
	Low cost	
Disadvantages	High machining accuracy requirement	Limited detection range
	Limited scanning angle	Flash may hurt human eye
	Poor signal-to-noise ratio	

the low point cloud density, the LiDAR with low line number is prone to bring the problem of low resolution. At present, the mechanical LiDAR is mainly divided into 16 lines, 32 lines, 64 lines, and 128 lines. The larger number of lines leads to the difficulty of the technique. Table 3.5 lists the advantages and disadvantages of mechanical LiDAR.

- **Solid-state LiDAR** has no rotating components. It mainly relies on electronic components to control the laser emission angle. Technical schemes of solid-state LiDAR includes optical phased array (OPA) and flash, as described below.
  1. OPA solid-state LiDAR uses multiple light sources to form an array. By controlling the luminescence time difference of each light source, the main beam with a specific direction can be synthesized, which can perform scanning in different directions.
  2. Flash solid-state LiDAR is a non-scanning LiDAR, which directly emits a large area of laser in a short time to cover the detection area and then completes the rendering of the image with a highly sensitive receiver. The wavelength of the emitted laser is a key factor. The advantages and disadvantages of solid-state LiDAR are shown in Table 3.6.

Table 3.7 Advantages and Disadvantages of Hybrid Solid-State LiDAR.

Advantages	Low cost
	High accuracy
	High stability
Disadvantages	Laser scanning is limited by the area of the MEMS scanning mirror
	Optical path is complicated

- **Hybrid solid-state LiDAR** has no mechanical rotating part. It uses micro-electro-mechanical system (MEMS) scanning mirror to change the transmitting angle of a single transmitter internally for array scanning. MEMS scanning mirror is a silicon-based semiconductor component in which all mechanical components are integrated into a single MEMS chip. Hybrid solid-state LiDAR belongs to the solid-state electronic component and still needs to use the mechanical way, that is, the MEMS scanning mirror swings up and down to carry on the scanning process. It is a compromise between mechanical and solid-state LiDAR; so it is defined as hybrid solid-state LiDAR. The advantages and disadvantages of hybrid solid-state LiDAR are shown in Table 3.7.

As the main part of the current unmanned vehicle to explore the surrounding environment, LiDAR's working principle is similar to that of ordinary radar, which is accepting multiple laser beams to establish a three-dimensional point cloud map. Accurate three-dimensional images are generated by photoelectric processing to realize the perception of velocity and position information. The working principle of the vehicle LiDAR system is shown in Figure 3.14. The laser ranging sensor working in the mode of pulse ranging is mounted on the autopilot platform. During the operation, the vehicle controls the rotation of the mechanical scanning structure through the control system and drives the ranging sensor to carry out rotational detection and ranging. According to the principle of pulse time-of-flight ranging, the ranging sensor measures the distance between the vehicle and the surrounding environment according to the specific working frequency. The distance information of the measured environment is transmitted to the computer system of the vehicle platform through mechanical scanning structure for data processing.

The on-board LiDAR adopts the principle of pulse time-of-flight ranging, which uses the accurate measurement of the time difference between the launch end and the target to realize distance calculation. Using laser as the signal source, the pulse laser emitted by the laser device hits trees, roads,

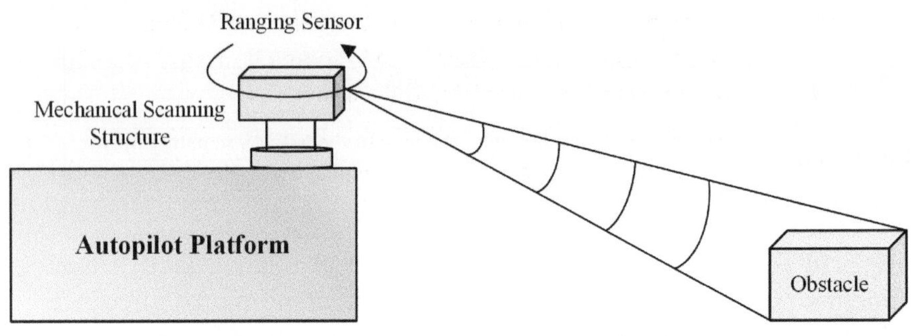

Figure 3.14 Working principle of on-board LiDAR system.

bridges, and buildings to cause scattering, and a part of the light wave will be reflected to the receiver of the LiDAR.

According to the ranging principle, the distance from the LiDAR to the target point is obtained. Let the flight time interval of the laser pulse be Δt, the target distance be R, and the propagation speed of the light in the air be c. Then, the ranging formula is shown as follows:

$$R = c \cdot \Delta t/2. \tag{3.10}$$

The mechanical rotating LiDAR currently in use generally has the following advantages and constraints.

- **Advantages of LiDAR:**
  1. High resolution and high ranging accuracy: LiDAR works in optical band, and its frequency is much higher than that of microwave. Therefore, compared with microwave radar, LiDAR has extremely high range resolution, angular resolution, and velocity resolution. It can completely draw the outline of the object and accurately measure the environment around the vehicle.
  2. Strong ability to resist active interference: The aperture of the LiDAR pulse beam transmitter is very small, that is, the receiver area is very narrow; so the chance of interference by other radar beams is very small.
  3. The distance, angle, reflection intensity, velocity, and other information of the target can be directly obtained to generate the multi-dimensional image of the target.
  4. LiDAR can conduct detection tasks without being affected by light. The detection effects are not affected by day or night and do not

depend on external lighting conditions or the radiation characteristics of the target itself. This is also an advantage that many camera sensors used in unmanned vehicles do not achieve.

5. With a wide range of speed measurement, LiDAR can successfully scan the contour of the obstacle with a relative speed of up to 200 kilometres per hour. For unmanned driving, LiDAR is not limited to urban areas or low-speed situations but can also be applied in high-speed mobile situations.

- **Constraints of LiDAR:**
  1. The high cost of LiDAR seriously restricts the commercial application of unmanned vehicles. Many mature technology schemes of unmanned driving use HDL-64E LiDAR, which is far more expensive than most of the models on the market.
  2. LiDAR is vulnerable to rain, snow, fog, and other adverse environments. This will also limit the operation of vehicles that rely on LiDAR for unmanned driving functions.
  3. LiDAR will produce massive data during operation, which poses a great challenge to the computing power of vehicle processors. In this case, if the on-board processor does not have enough computing capability, the command execution will stall or delay, thus threatening unmanned driving in complex road conditions.

3.5.3 Attack Technique Analysis of LiDAR

LiDAR has become a standard configuration for self-driving cars. According to the reflected signals, it can judge the position of the surrounding obstacles and their trajectory so as to establish a three-dimensional image of vehicle's surrounding environment. In this way, the self-driving car can use the map drawn by LiDAR to move freely. In addition, LiDAR has the advantages of high resolution, high ranging accuracy, strong ability to resist active interference, good detection performance, large speed range, and so on. Therefore, LiDAR is a necessary part for vehicles to obtain environment information. If the LiDAR fails, the decision of the unmanned driving system will be inaccurate.

3.5.3.1 Relay Attack

For replay attacks, the researchers recorded signals from a commercial ibeo LUX 3 LiDAR. Since laser signals from this LiDAR are not encoded and encrypted, hackers can replay signals at any time. The only trick to be noticed is to send the laser signal back to the LiDAR at the right time so as to create the illusion that the laser signal is sent back as it encounters an obstacle. When the self-driving car receives the false signal, it will consider that there is an obstacle.

Using replay attack, attackers can make a self-driving car consider that something is in front of it, forcing it to slow down. Also, they can use a large number of false signals to mislead the car and make it remain stationary because it does not tend to hit potential obstacles.

3.5.3.2 Spoofing Attack

For spoofing attack, researchers can simulate the signal reflection of several obstacles, such as cars, walls, pedestrians, and so on. At the same time, they can clone many signal backups, resulting in the illusion that obstacles are moving. In addition, the tracking system of LiDAR can be subjected to denial of service attack. In other words, the tracking system of LiDAR is disabled and it cannot emit laser signals or receive reflected signals; thus, the LiDAR sensor cannot track the real obstacles.

3.6 Experiments

In this chapter, we will show a variety of attack experiments on unmanned vehicles, including attacks on ultrasonic radar, millimeter wave radar, hi-definition camera, and LiDAR, all of which are sensors that driverless cars rely on. These attacks can have serious impacts on the normal and safe driving of unmanned vehicles and are worthy of attention.

3.6.1 Attack Experiment Against Ultrasonic Radar

Researchers from Zhejiang University and Qihoo 360 SKY-GO vehicle cyber security team published paper "Can You Trust Autonomous Vehicles: Contactless Attacks against Sensors of Self-driving Vehicle" in DEFCON 2016 [11]. They described the vulnerabilities of ultrasonic radar, millimeter wave radar, high-definition camera, and so on for vehicles with autonomous driving capabilities. An analysis of these studies and how vulnerabilities or flaws can be exploited to attack unmanned vehicles will be presented in this chapter.

3.6.1.1 Jamming Attack

The principle of jamming attack is to continuously transmit ultrasonic wave to the on-board ultrasonic radar to reduce signal-to-noise ratio. First, on-board ultrasonic sensors typically operate at frequencies between 40 and 50 KHz (closer to 50 KHz). In order to make the transducer show the best emissivity and sensitivity, it is better to choose the interference sensor in the same frequency band as the on-board ultrasonic radar. However, it is very difficult to get 50-KHz transducer on the market. Fortunately, experiments have shown that the 40-KHz sensor also works. Researchers applied low-cost Arduino UNO boards to generate controllable square waves at 40 KHz. Meanwhile, a higher voltage was applied to realize further attack distance. For this purpose, the researchers used Arduino's 5-V output, which worked well within a limited range.

The researchers tested cars with ultrasonic radars and self-driving features from Audi, Volkswagen, Ford, and Tesla both indoors and outdoors and further tested the Tesla Model S's AP and summoning capabilities. In the experiment, the obstacle used for testing can be normally detected by the sensor before attack.

The researchers first carried out attacks on eight different ultrasonic radars in laboratory, including ultrasonic ranging modules, aftermarket vehicle sensors, as well as parking assistance systems. In the experiment, the jamming attack obtains two output results, that is, showing zero distance and maximum distance on the dashboard. The former means that the ultrasonic radar detects obstacles that are very close to it, while the latter means that no obstacles are detected. These two results are caused by different intrinsic designs of the sensor. In the first case, the ultrasonic sensor sets a certain threshold for the detection echo. If the interference signal in the experiment always exceeds the threshold, it will be incorrectly identified as the returned echo; so the dashboard will show the distance as zero. In another case, the ultrasonic radar designed a flexible threshold for detecting the echo. The interference signal in this case is considered to be noise, which will reduce the signal-to-noise ratio, and, therefore, the dashboard will show the maximum distance.

In addition, the researchers tested four cars with unmanned driving functions. As shown in Figure 3.15, the ultrasonic jammer A is placed in front of the bumper, and B represents the ultrasonic radars on Tesla Model S. As shown in Figure 3.16, ultrasonic radar can work properly without attack. When a jamming attack is conducted, the vehicle will not be able to detect the obstacle, that is, the distance detected is the maximum, and the vehicle will not alert the driver. The researchers further tested the situation when the

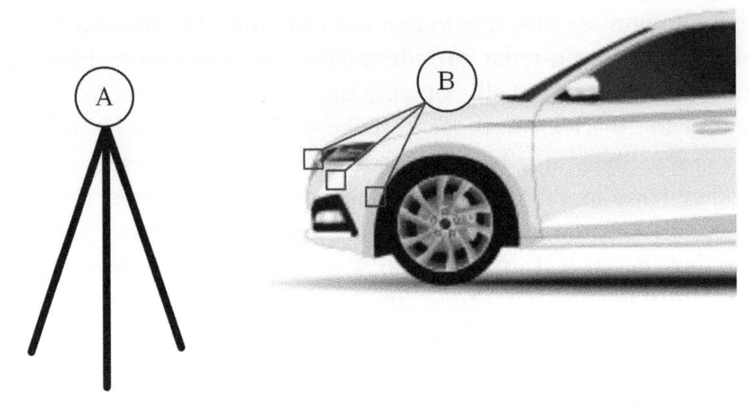

Figure 3.15 Experiment settings of attacks against ultrasonic radar.

Figure 3.16 Jamming and spoofing attack results on Tesla Model S [11].

car was running in reverse, and the results of the attack experiment were the same.

3.6.1.2 Spoofing Attack

Spoofing attack is to use the generated ultrasonic pulse to deceive the ultrasonic radar, making it mistake the false pulse as the echo from the obstacle. In this attack, false pulse needs to reach the sensor before the true echo. By adjusting the trigger time of this pulse, a malicious attacker can manipulate the measurement distance of the ultrasonic radar. Experimental settings for this attack are similar to the aforementioned jamming attack. To deceive the ultrasonic radar, the excitation time of 200–300 μs worked well, and the researchers suggested it to be no more than 1 ms.

It is worth noting that the on-board ultrasonic radar only detects the nearest obstacle. Only the first reasonable echo will be received by the radar, and the other echoes will be ignored. Therefore, to ensure the success of the attack, the false echo must reach the ultrasonic radar before the real echo. That is, the spurious signal must be injected into the time slot between the end of the transmitting pulse and the beginning of the first detected echo. The length of this time is determined by the distance of the ultrasonic radar to the obstacle.

Therefore, the injection time as well as the length and period of the forged echo all affect the attack. Also, a set of effective sensor output results can be found through repeated experiments. As shown in Figure 3.16, when there are no obstacles, spoofing attacks can cause false obstacles to be displayed within the vehicle's detection range. This can further lead to misjudgement by drivers and wrong decisions by self-driving cars.

When ultrasonic radar is jammed or spoofed, unmanned vehicles which rely heavily on ultrasonic radar cannot effectively detect encountered obstacles. That is, obstacles are detected where there are no obstacles, or obstacles are not detected where there are obstacles. In these cases, unmanned driving systems cannot carry out accurate and real-time path planning and navigation. It may cause vehicles to deviate from a good route or even lead to serious traffic accidents, resulting in vehicle damage and driver casualties.

3.6.2 Attack Experiment Against Millimeter Wave Radar

3.6.2.1 Jamming Attack

Jamming attack requires some technical details of the millimeter wave radar used on the Tesla Model S, which have not been disclosed. Therefore, attackers can analyse technical details through manufacturers and search for information about the vehicle. On the other hand, millimeter wave radar's spectrum and waveform can be directly observed. However, this method is not easy. Tesla Model S installed 76–77 GHz millimeter wave radar produced by Bosch company. After confirming the frequency band, equipment that works in this band can be used to observe the waveform of the millimeter wave. Common spectrum analyser and signal generator are difficult to achieve such high frequencies. In addition, instruments that can reach high frequency are usually very expensive. At this point, frequency multipliers and mixers can be used to achieve the increase of frequency.

Researchers used DSOS804A oscilloscope, Keysight N9040B UXA analyser, VDI 100 GHz mixer, and 89601B VSA software for signal analysis.

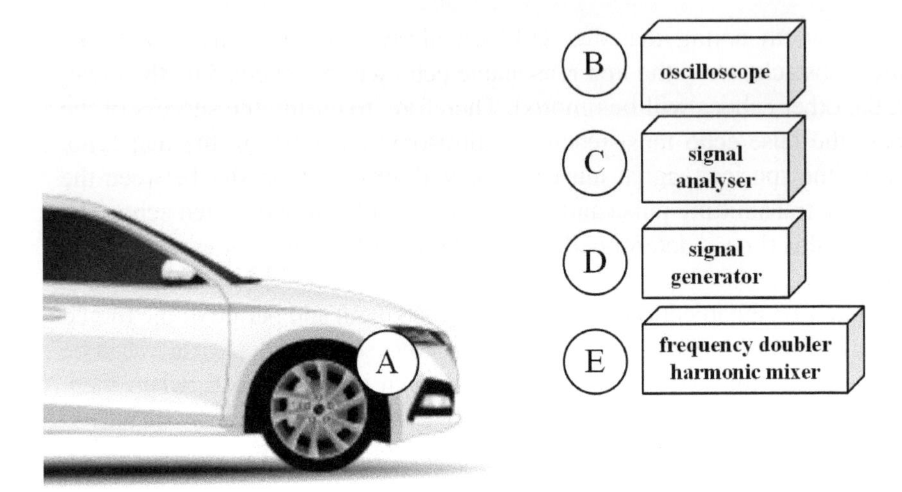

Figure 3.17 Experiment settings of attacks against millimeter wave radar.

Figure 3.17 shows the experimental settings for jamming attacks on millimeter wave radar, in which A is Tesla's millimeter wave radar, B is a high-definition oscilloscope, C is a signal analyser, and D is a signal generator. The frequency doubler and harmonic mixer are placed at E. The researchers placed the antenna half a meter away from the car, and its horizontal direction is consistent with the on-board millimeter wave radar. When Tesla's millimeter wave radar starts, the driver can see the millimeter wave radar's working condition on the dashboard. Using a signal analyser, the researchers confirmed that millimeter wave radars can operate at 76–77 GHz. After manual correction, they determined that the operating bandwidth of the millimeter wave radar was about 450 MHz with the FMCW modulation.

The principle of jamming attack is to transmit signals to interfere with millimeter wave radar within the same frequency bandwidth. There are two options for jamming waveforms, fixed frequency at 76.65 GHz or sweep within 450 MHz bandwidth. The millimeter wave system may treat the interference signal sent by the attacker as a strong noise or error signal, resulting in signal-to-noise ratio reduction or computation error. This will further lead to the failure of millimeter wave radar. The attack effect is shown in Figure 3.18. The left figure is the normal running state of the vehicle on the dashboard. The middle figure is the state of unmanned driving when the millimeter wave radar can detect the car in front of it normally. The right

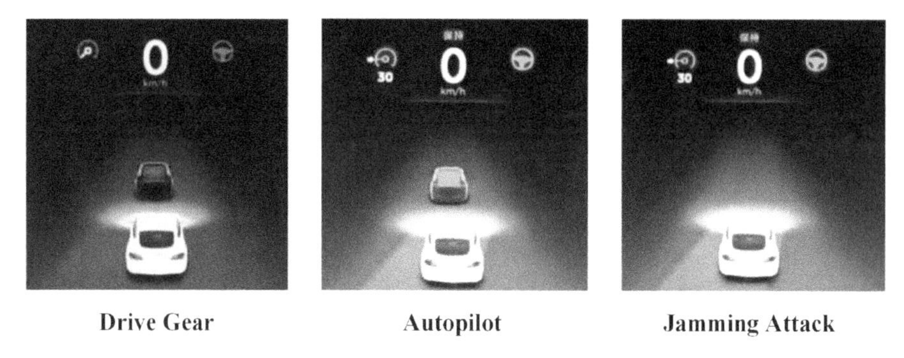

Figure 3.18 Jamming attack results on Tesla Model S [11].

figure shows the state after jamming attack, and, as a consequence, the car that should be detected in front of the vehicle disappears.

Jamming attacks can cause obstacles in the vehicle detection range to suddenly disappear by applying signals in the same frequency band as millimeter wave radar. This is a serious threat to self-driving cars. It will not only interfere with the judgement of the driver in low-level unmanned driving but also cause the unmanned system to make wrong decisions and construct wrong paths.

3.6.3 Attack Experiment Against Hi-Definition Camera

3.6.3.1 Blinding Attack

Data from a variety of radars and many other sensors are insufficient to achieve safe unmanned driving, especially in complex urban scenarios. Camera can recognize the surrounding environment of the vehicle in the form of images, obtain road signs and road information clearly, and help the unmanned driving system to better control the vehicle behaviour. Camera sensor plays an important role in driverless perception systems. Hi-definition cameras are mainly used in the following aspects: road detection, vehicle detection, pedestrian identification, as well as traffic sign detection. Camera sensors have advantages such as the information obtained is very intuitive, and it has high recognition accuracy. However, camera relies on external light sources to work. Low illumination, direct sunlight, incoming and outgoing tunnel, rain, snow, and haze can reduce camera sensitivity or even cause a malfunction.

The attacks on camera sensor are based on the fact that the camera may be interfered by malicious light source, which can produce hard-to-recognize

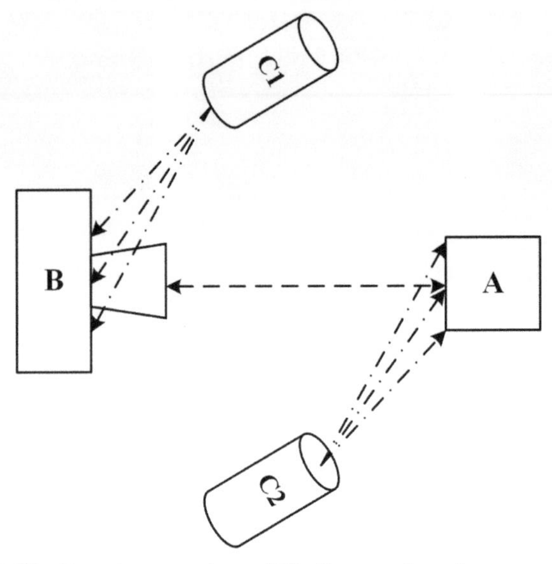

Figure 3.19 Experiment settings of blinding attack against camera sensor.

images, thus affecting the decision making and control of the unmanned driving system, and even leading to serious traffic accidents. Photoelectric sensors are very sensitive to light intensity. Most of the laser energy can be absorbed between the absorption coefficients 10^3 and 10^5. Under laser irradiation, the thermal effect caused by the inhomogeneous temperature field will cause the surface temperature of the camera to rise rapidly. This can cause permanent damage to the camera's optoelectronic components. Three types of light sources might be used in this kind of attack, namely LED light, laser light, and infrared LED light. Figure 3.19 shows the experimental environment of blinding attack. The calibration board A is located 1 m away from camera B. Light sources C1 and C2 are used to direct camera or calibration panel.

The method of blinding attack is to use the light source to direct the camera or to direct the calibration board so that the reflection is directed to the camera. The final effect depends on the distance between the light source and the camera as well as the intensity of the light source. By using LED to illuminate calibration board, the tonal value of the central area will increase, and the information in this area can be completely hidden so that the camera cannot recognize it, as shown in the left part of the first row in Figure 3.20. Moreover, a direct illumination of the camera can result in a significantly

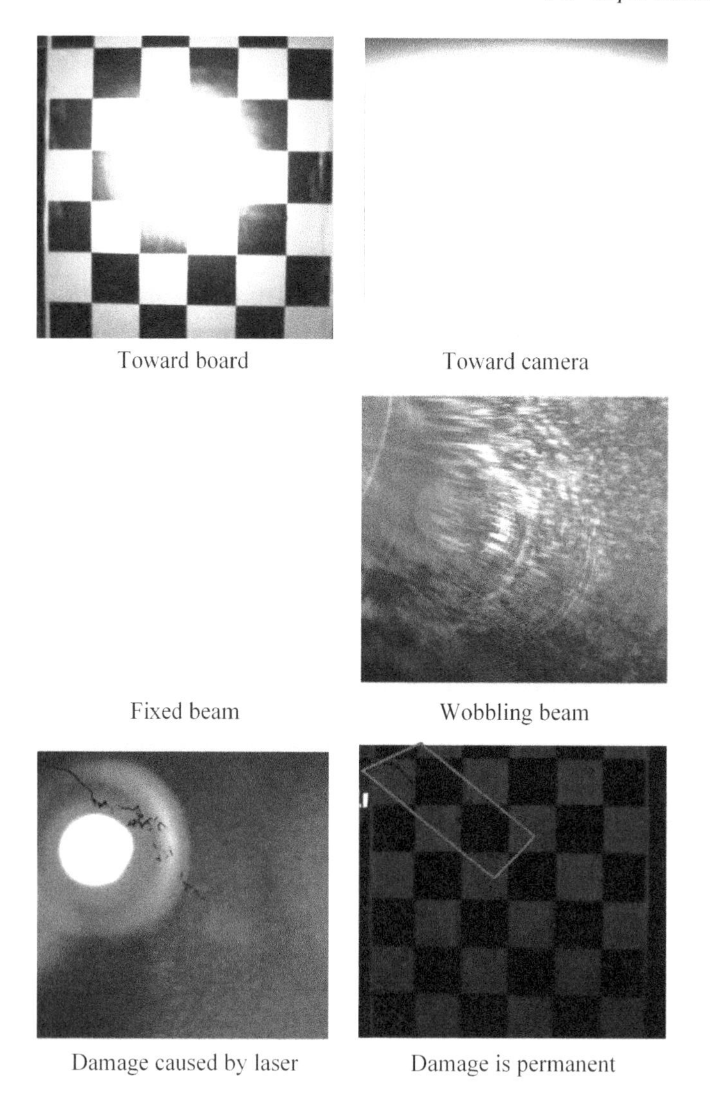

Figure 3.20 Blinding attack against camera sensor with LED spot and laser [11].

higher tonal value, resulting in a complete failure of the camera, as shown in the right part of the first row in Figure 3.20. If the distance increases, the effect decreases. If the light source is too strong, it will cause the camera to burn out directly.

The laser beam against the calibration board will not cause damage to the camera, but shooting the camera directly will cause it to fail for a few seconds. At the same time, the use of wobbling light sources may also lead to a brief failure of the camera as shown in the right part of the second row in Figure 3.20. To permanently damage the camera chip, attackers can use the laser to direct the camera at close range for a few seconds. As shown in the bottom of Figure 3.20, black curve appears. In addition, researchers found that infrared LED light cannot damage the camera.

Camera is an important visual sensor for unmanned vehicles. If it is attacked and the output results are inaccurate, or its hardware is permanently damaged, then the vehicle's ability to detect the surrounding environment will be seriously affected, thus affecting the normal functions of the decision making module and navigation module of the unmanned vehicle system.

3.6.4 Attack Experiment Against LiDAR

3.6.4.1 Relay Attack

Jonathan Petit *et al.* from Security Innovation published a paper "Self Driving and Connected Cars Fooling Sensors and Tracking Drivers" in the security conference BlackHat Europe in 2015 [7]. They conducted attacks against LiDAR. An analysis of these studies and how the vulnerabilities can be exploited to impact unmanned vehicles are presented in this chapter.

The LiDAR tested by researchers is ibeo LUX 3, which is a four-layer LiDAR. It cannot provide three-dimensional view and can only provide four-layer two-dimensional plane view. This kind of LiDAR uses the following techniques.

- Double output technology: By using power saving technology, the measuring distance can exceed 200 m. Even under severe weather conditions, it can maintain high sustainability and reliability.
- Multiple echo technology: By using the optimized measurement technology for the target object, the number of echoes for target detection is increased, that is, multiple echoes are processed during each measurement so that the measured object can be reliably restored by the measurement results.
- Angle resolution technology: Three angle resolutions are available, which can be adjusted according to different application scenarios to obtain better detection results. For unimportant occasions, the angle resolution can be reduced appropriately.

The relay attack is designed to relay the signals of the on-board LiDAR, produce false echoes, and eventually deflect the actual obstacles from their

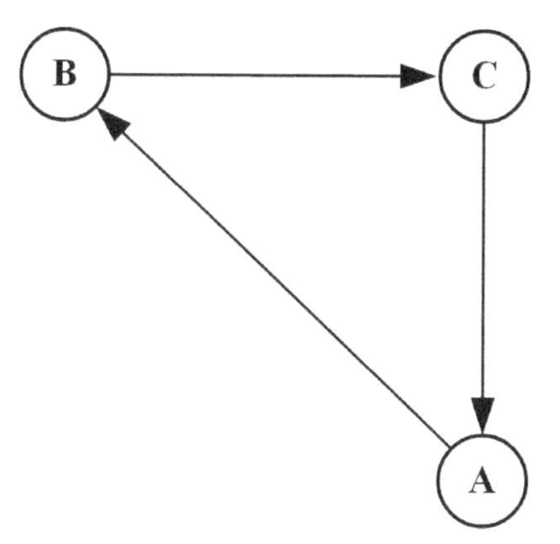

Figure 3.21 Experiment settings of relay attack against LiDAR.

positions. Two signal transceivers are required to conduct a relay attack. As shown in Figure 3.21, the transceiver B is a photoelectric detector, which outputs the voltage signal corresponding to the pulse intensity of LiDAR A. B's output signal will be received by C, and then C will send another pulse as feedback.

Also shown in Figure 3.21, in this experiment, the two transceivers are 1 m apart. In practice, a roadside hacker can conduct relay attacks by receiving the LiDAR signal from a vehicle and forwarding it to the target vehicle in another location.

The effect of relay attack on LiDAR's environmental sensing performance is shown in Figure 3.22. Before relay attack, the LiDAR detected only a

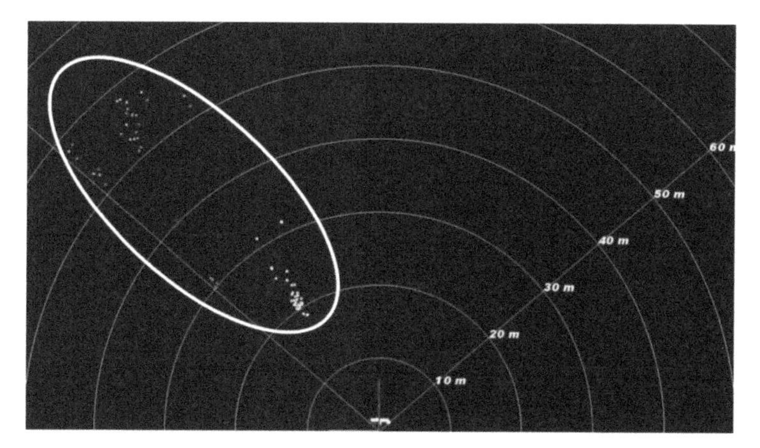

Figure 3.22 Experiment results of relay attack against LiDAR [7].

wall in front of it, that is a yellow line at the bottom of the figure. As shown in the white circle in Figure 3.22, when the relay attack is applied, the LiDAR will receive the echo of distant object, which is actually forged by the attacker using relay method. Since the unmanned driving system can detect obstacles at a long distance, these echoes will seriously affect the path planning function of the vehicle. This experimental result indicates that the laser signal is not encoded and can be used by attackers to carry out relay attacks to produce false echo and deceive the LiDAR.

3.6.4.2 Spoofing Attack

The relay attack on LiDAR indicates that the environment in which the LiDAR is located can easily be injected with false echo. Subsequently, further experiments can be performed by creating false objects. This can be achieved by using the original signal as the trigger signal to actively spoof the LiDAR.

The maximum detection distance of the ibeo LUX 3 LiDAR used in this experiment is 200 m, and the signal fluctuates in about 1.33 μs. The premise of the success of this experiment is to inject the false signal into the LiDAR in this time slot. The time when the LiDAR receives the signal determines its distance from the obstacle. Therefore, hackers can use this method to control the location of forged obstacles.

As shown in Figure 3.23, after receiving the first echo, the LiDAR received the forged pulse echo. This will deceive the LiDAR to consider that the pulse signal goes further, making the obstacle look farther. If the LiDAR receives a forged pulse in a silent window after 1.33 μs, the spoofing attack will not succeed.

The experimental settings are shown in Figure 3.24, in which A is the LiDAR used for spoofing attacks, B is the transceiver, and P1 and P2 are the

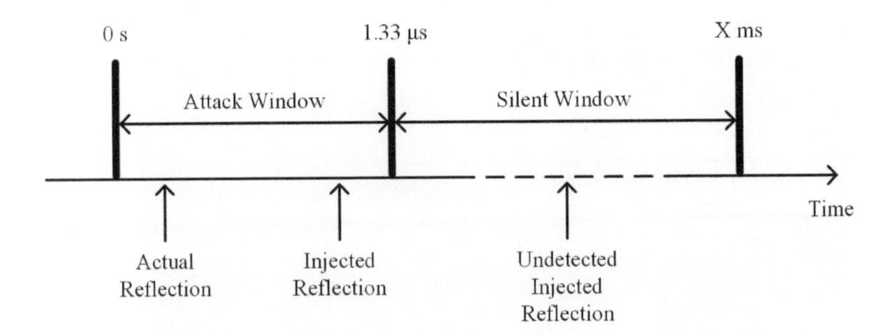

Figure 3.23 Attack window of spoofing attack against LiDAR.

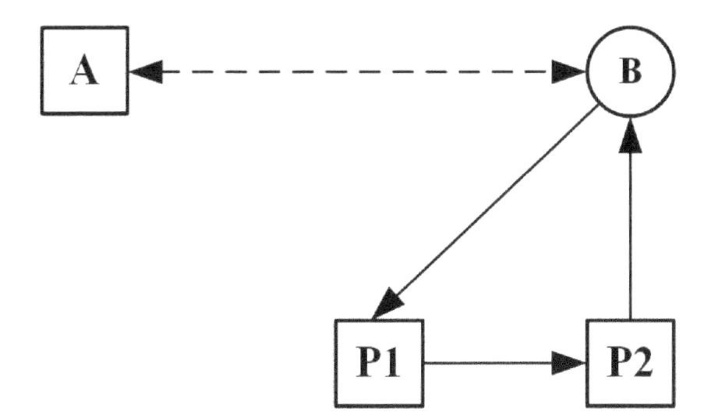

Figure 3.24 Experiment settings of spoofing attack against LiDAR.

control logic units. In this experiment, the pseudo-signal is generated by the control logic module. In addition, the output of the transceiver is linked to the input of P1, and P1 delays its own output. When the output of one pulse generator P1 is linked to the other pulse generator P2, it will cause P2 to generate square pulses, which, in turn, will be transmitted to B.

In this experiment, the delay of the forged pulse and the number, width and period of the pulse are controllable factors. Figure 3.25 presents the effect of delayed output and pulse replication on spurious signals in spoofing attack. When the control logic is triggered, a fixed number of similar pulses will be generated. At this point, the pseudo-signal can be very similar to the original signal by using an oscilloscope to adjust its parameters.

The effects of spoofing attack are shown in Figure 3.26. Among them, the left figure presents that the LiDAR detected forged wall copies at a long distance. By adjusting the delay time of the forged reflection echo, the wall can look closer or farther until the signal falls outside the attack window. The pulse generator might be set to output a number of pulses when triggered; in this way, the spoofing attack can also forge multiple false echoes continuously to create multiple obstacles. For example, the forged wall signal is shown in the right figure.

Figure 3.27 shows the LiDAR's scanning results. It can be seen from the figure that in a short period of time, the second forged wall copy is classified as three new obstacles; so that the LiDAR cannot track the obstacle continuously. Therefore, forged objects can be used to deceive the ibeo LUX 3 LiDAR as well as affect the normal operation of LiDAR and the unmanned vehicle.

The difficulty of attacking or cracking an unmanned vehicle is highly dependent on vehicle's configuration. Single-point attacks are easy, but they

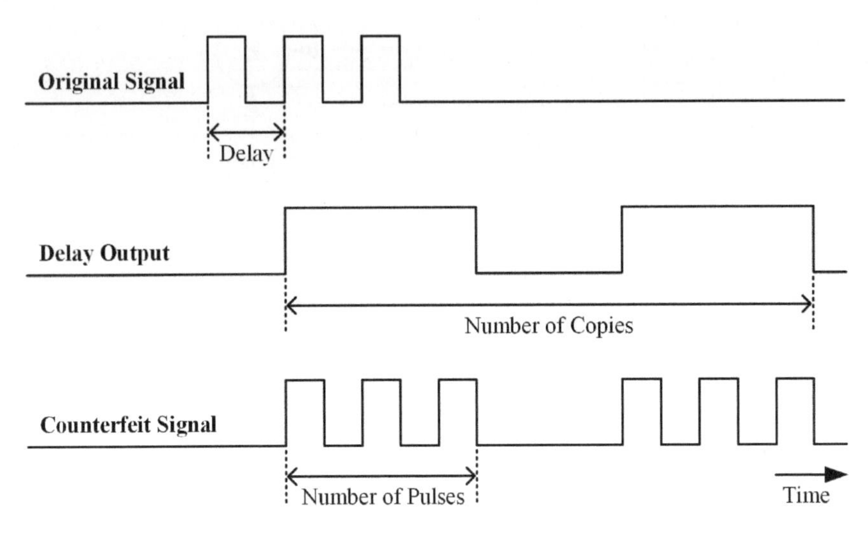

Figure 3.25 The controllable factors of spoofing attack.

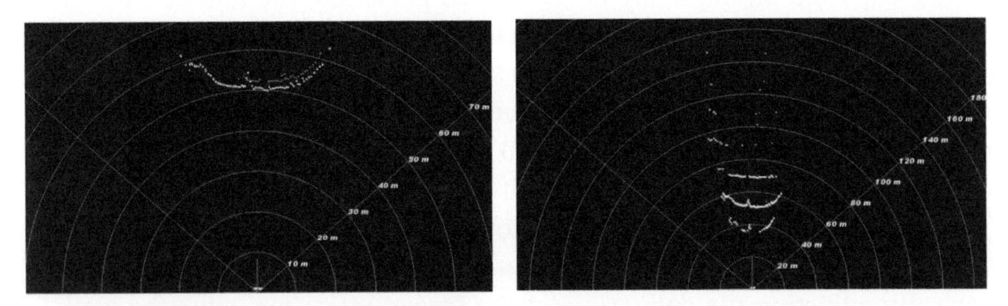

Figure 3.26 Spoofing LiDAR with one copy and multiple copies [7].

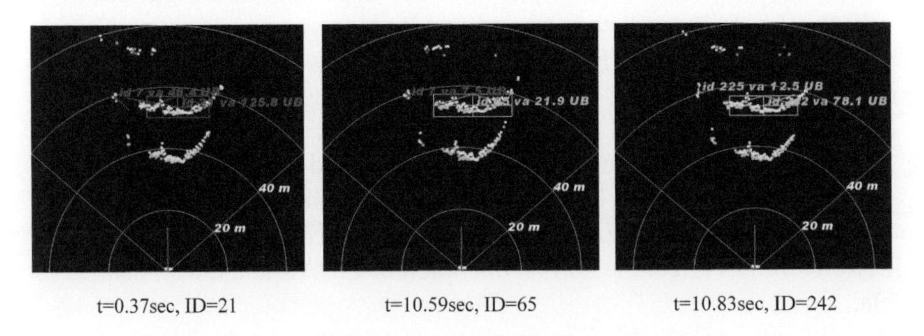

Figure 3.27 LiDAR scanning results of forged wall tracking [7].

only work on one vehicle module. If all parts of an unmanned car fit together perfectly, the impact of the attack will be reduced. Therefore, although LiDAR has obvious vulnerabilities, they can be complemented by cameras, millimeter wave radars, or positioning systems. How to break through multi-layer sensor systems still requires researchers to keep trying.

Bibliography

[1] The attack vector "bluebrone" exposes almost every connected device. https://www.armis.com/blueborne/. [Online; accessed Jan. 2021].

[2] Andrew P Bradley. The use of the area under the ROC curve in the evaluation of machine learning algorithms. *Pattern recognition*, 30(7):1145–1159, 1997.

[3] Yurui Cao, Qian Luo, and Jiajia Liu. Road navigation system attacks: A case on GPS navigation map. In *IEEE International Conference on Communications (ICC)*, pages 1–5, 2019.

[4] Tom Fawcett. An introduction to ROC analysis. *Pattern recognition letters*, 27(8):861–874, 2006.

[5] Earlence Fernandes, Bruno Crispo, and Mauro Conti. Fm 99.9, radio virus: Exploiting FM radio broadcasts for malware deployment. *IEEE Transactions on Information Forensics and Security*, 8(6):1027–1037, 2013.

[6] Yansong Li, Qian Luo, Jiajia Liu, Hongzhi Guo, and Nei Kato. TSP security in intelligent and connected vehicles: Challenges and solutions. *IEEE Wireless Communications*, 26(3):125–131, 2019.

[7] Jonathan Petit, Bas Stottelaar, Michael Feiri, and Frank Kargl. Remote attacks on automated vehicles sensors: Experiments on camera and lidar. *Black Hat Europe*, 11, 2015.

[8] Ishtiaq Rouf, Robert D Miller, Hossen A Mustafa, Travis Taylor, Sangho Oh, Wenyuan Xu, Marco Gruteser, Wade Trappe, and Ivan Seskar. Security and privacy vulnerabilities of in-car wireless networks: A tire pressure monitoring system case study. *USENIX Security Symposium*, 10, 2010.

[9] Yijie Xun, Jiajia Liu, Jing Ning, and Haibin Zhang. An experimental study towards the in-vehicle network of intelligent and connected vehicles. In *IEEE Global Communications Conference (GLOBECOM)*, pages 1–6, 2018.

[10] Yijie Xun, Jiajia Liu, and Zhenjiang Shi. Multi-task learning assisted driver identity authentication and driving behavior evaluation. *IEEE Transactions on Industrial Informatics*, 2020. DOI: 10.1109/TII.2020.3034276.

[11] Chen Yan, Wenyuan Xu, and Jianhao Liu. Can you trust autonomous vehicles: Contactless attacks against sensors of self-driving vehicle. *DEF CON*, 24, 2016.

Index

A
Advanced Driving Assistance System (ADAS), 12, 135

C
Connected and Autonomous Vehicle (CAV), xi
Controller Area Network (CAN), 1, 67, 144
Cryptography, 81
C-V2X, xi
Cyber Security, 152

E
Electronic Control Unit (ECU), 69, 131

I
Intelligent and Connected Vehicle (ICV), xi, 2, 167, 170
Intra-vehicle Communication, 53
In-vehicle Network, 8, 12, 44, 167

O
On-Board Diagnostic II (OBD-II), 24

P
Privacy, 29, 74, 83, 106

S
Security, 1, 28, 53, 58
Standardization, 2, 24

U
Unmanned Driving, 117, 118, 129, 132

V
Vehicle Bus, 1, 9, 12, 35
Vehicle to Everything (V2X), xxiv, 122

Index

About the Authors

Jiajia Liu is a full professor (Vice Dean) with the School of Cybersecurity, Northwestern Polytechnical University. He has published more than 180 peer-reviewed papers in many high-quality publications, including prestigious IEEE journals and conferences. He received IEEE VTS Early Career Award in 2019, IEEE ComSoc Asia-Pacific Outstanding Young Researcher Award in 2017, IEEE ComSoc Asia-Pacific Outstanding Paper Award in 2019, Niwa Yasujiro Outstanding Paper Award in 2012, the Best Paper Awards from many international conferences including IEEE flagship events, such as IEEE GLOBECOM in 2016 and 2019, IEEE WCNC in 2012 and 2014, IEEE WiMob in 2019, IEEE IC-NIDC in 2018. His research interests cover a wide range of areas including intelligent and connected vehicles, wireless and mobile ad hoc networks, and Internet of things security. He has been actively joining the society activities, like serving as associate editors for IEEE Transactions on Wireless Communications, IEEE Transactions on Computers and IEEE Transactions on Vehicular Technology, editor for IEEE Network, editor for IEEE Transactions on Cognitive Communications and Networking, and serving as technical program committees of numerous international conferences. He is the Vice Chair of IEEE IOT-AHSN TC, and is a Distinguished Lecturer of IEEE Communications Society and Vehicular Technology Society.

Abderrahim Benslimane is Full Professor of Computer-Science at the Avignon University/France since 2001. He has the French award for Doctoral supervision and Research 2017-2021. He has been recently an International Expert at the French Ministry of Foreign and European affairs (2012-2016). He served as a coordinator of the Faculty of Engineering and head of the Research Center in Informatics at the French University in Egypt. He was attributed the French award of Scientific Excellency (2011-2014). He is past Chair of the ComSoc Technical Committee of Communication and Information Security. He is EiC of Inderscience Int. J. of Multimedia Intelligence and Security, Area Editor of Security in IEEE IoT journal, Area Editor of Wiley Security and Privacy journal and editorial member of IEEE Wireless Communication Magazine, IEEE System Journal, Springer Wireless Network Journal and Elsevier Ad Hoc. He is founder and serves as General-Chair of the IEEE WiMob since 2005 and of iCOST and MoWNet international conference since 2011. He is General Chair of IEEE CNS 2020, Executive Forum Co-Chair at IEEE Globecom 2020, Program vice Chair of IEEE TrustCom 2020, and Program Chair of IEEE iThings 2020. He has more than 210 refereed international publications and 14 Special issues.